MILITARY INDUSTRY IN TAIWAN AND SOUTH KOREA

MILITARY INDUSTRY IN TAIWAN AND SOUTH KOREA

JANNE E. NOLAN

St. Martin's Press   New York

First published in the United States of America in 1986

Printed in Hong Kong

ISBN 0–312–53234–2

Library of Congress Cataloging-in-Publication Data
Nolan, Janne E.
Military industry in Taiwan and South Korea.
Bibliography: p.
Includes index.
1. Munitions – Taiwan.   2. Munitions – Korea (South)
I. Title.
HD9743.T282N65 1985   338.4'76234'0951249   85-22157
ISBN 0–312–53234–2

To Barry

Contents

List of Figures

Acknowledgements

This book is a reworked version of a doctoral thesis presented in December 1982 to the faculty of the Fletcher School of Law and Diplomacy. In addition to my gratitude for the financial support provided by the Fletcher School and its International Security Studies Program, I owe an extraordinary intellectual debt to many members of Fletcher's faculty. In particular, I should like to thank Geoffrey Kemp, who encouraged me to pursue a less than conventional research topic that had received little attention from experts and was considered controversial. Scott Thompson's cogent criticism of drafts and his insistence on adherence to deadlines permitted the original study to emerge as a reality rather than remain an elaborate conceptual fantasy. Phyllis Webber, the indefatigable and patient source of administrative information at Fletcher, provided support and encouragement throughout.

It is a special tribute to the Fletcher School that it affords its students the flexibility to expand their experience outside the academic setting prior to completing the doctoral degree. In my case, I found that several years spent in government service helped immeasurably to broaden my understanding and appreciation of international security policy and the complex subjects analyzed here. I am indebted to my colleagues and mentors in the Arms Control and Disarmament Agency for contributing to a challenging professional experience during these 'sabbatical' years.

The opportunity to join the Stanford Arms Control Program in 1980 was a stroke of good fortune for which I am very grateful. Aside from providing a generous research grant and an extraordinarily pleasant professional environment, Stanford gave me a year of invaluable exposure to a rich and diverse intellectual community and to a number of colleagues who I have come to value as close friends. I owe special thanks to John Lewis, both for bringing me to Stanford and for encouraging my participation in an eclectic array of conferences and seminars that added to my intellectual perspective. Chip Blacker deserves more gratitude than can properly be expressed in this context;

he is a permanent source of inspiration. This is equally true for Matthew State and Barbara Johnson, who kept me sane and entertained. I should also like to thank my fellow Fellows, Cindy Roberts, Gloria Duffy and Condi Rice, for providing a challenging and convivial atmosphere in which to work. Ed Laurence, Alex George and Sid Drell also contributed to the personal success of my year at Stanford.

I completed my dissertation in Washington in 1981–2 under the auspices of the Georgetown Center for Strategic and International Studies, which kindly provided me with office space and a professional environment in which to work. For this, I owe special thanks to Mike Moodie and Amos Jordan, and to members of the Soviet Project, whose suite I occupied for several months. Among the latter, special gratitude goes to Jim Townsend, and to Aileen Masterson-Moodie for her ingenuity and hospitality, as well as for being a first-rate colleague. The Institute for the Study of World Politics granted me financial support during this period.

It is during overseas travel that one is most grateful for assistance. I owe my initial opportunity to travel to Taiwan to Ed Luttwak. Tom Robinson was extraordinarily helpful to me in setting up interviews in Asia. In Taiwan, I was graciously hosted by the Freedom Council, from which I acquired introductions to many individuals whose knowledge and perspective were invaluable to my research. Individuals at the American Institute in Taiwan also gave generously of their time, both in Taipei and in Washington; I am especially indebted to Larry Ropka. In Korea, extensive interviews were arranged for me by individuals at the US Embassy, particularly Kevin Kearns.

American officials in both Korea and Taiwan were gracious and helpful far beyond the call of duty. Special mention is reserved for the Joint US Military Advisory Group in Korea, including Colonel Vargo, Captain Perkins and Major Mills, who arranged special briefings on my behalf and also provided important insights for this study from their own experience.

The help of the Ryu family was critical to the success of my visit to Korea. It was through the kindness of Mr Chang Ryu, the personnel of his firm (especially Mr Kim in New York) and, most especially, his son Chung, that I was able to visit Korean defense plants and gain direct knowledge of their workings. Chung served as interpreter, guide and friend throughout my visit.

I should like to thank the individuals of the Northrop Corporation who gave of their time: Rob Silverstein, Bob Wachter and Fritzy

Ermarth, as well as the many members of the US policy community who shared their confidences with me.

There is a large debt of gratitude left for a number of friends without whose intellectual and emotional support I could not have completed this work. Jim Fluhr was a loyal and patient friend from the beginning, offering helpful advice on my research design and serving as an important source of information and guidance. Myrtle Ramsey's alchemy turned mangled drafts into impeccable typescript, with much good sense and cheery optimism thrown into the bargain. Althea Duersten was an ever-loyal friend who, as friends ought, reminded me frequently of the ignominy that would descend were I to fail to complete my work. A good deal of academic usage indispensable to any respectable doctoral thesis but irksome to the lay reader was removed by Steven Kennedy. As a result, the text lost weight without losing strength. I owe special debts to James Nolan, whose financial support was the critical underpinning of this whole endeavor; to Margaret Claughton Nolan, without whom this project would literally not have been possible; and to Kenneth Ullman, whose wise counsel kept me on a reasonably even keel.

Finally, it is to Barry Blechman that I direct much praise and admiration. He is the true hero of this epic.

1 Introduction: Defense in the Dependent State

Since their creation as independent states, and throughout the period of rapid development of their industrial economies, Taiwan and South Korea have had close relations with the United States. These relations could be characterized as mutually dependent, although in an unbalanced way. While the two countries have depended on the United States for their protection – particularly crucial in the early years – and for massive infusions of economic and military aid, the United States in turn has expected these states to serve as outposts of American power in the postwar global order. Defined initially in the heyday of the *Pax Americana*, this mutual dependency persists, although it has been strained both by changes in great power relationships (the US rapprochement with the People's Republic of China) and by the maturation of what were once compliant client states.

In this study, we shall trace the progress of these relationships of mutual dependence, examining the struggle on the part of the leaders of Taiwan and South Korea to increase the autonomy of their states within a context of constant threat in which choices seem narrow indeed. We shall see that this yearning toward autonomy was expressed in the complementary decisions to acquire an independent defense capability and to build the heavy industrial base necessary to make indigenous defense production possible. We shall examine the defense production potential of Taiwan and South Korea in detail, based on an assessment of the economic, political and social factors that may serve to enhance or constrain this potential. We shall also consider the way in which the growth of a defense industrial sector affects key aspects of government policy in the military, political and economic spheres, so as to permit judgments about the extent to which various types of defense investment serve to promote or inhibit the achievement of overall security objectives.

1

WHY ARE TAIWAN AND SOUTH KOREA IMPORTANT?

Taiwan and South Korea have an historical legacy of almost exclusive dependence on the United States for their military – and therefore political – security. The United States has been willing to support these states because of what it perceives to be their strategic position in an area deemed to be of profound importance for its own larger sense of security. Nevertheless, since 1969 particularly, the United States has sought ways to shift some of the burden for its mutual security arrangements to its partners on the 'periphery' of its defense and trade network. Taiwan and South Korea have responded to this and to previous evidence of flagging US commitment by devoting growing increments of national resources to the development of independent defense industries and by undertaking far-ranging policies to achieve greater self-reliance in all spheres over the past two decades. They are now committed politically, militarily and economically to achieving success in these endeavors.

From a more academic point of view, Taiwan and South Korea are particularly interesting for a study of this kind because of the urgency of the security threats they both face, an urgency that permeates all aspects of national life. As such, defense considerations affect national planning at virtually every level and in every sphere, providing a rich environment in which to observe the interaction of defense and development variables. Their similarities allow us to assess how countries in similar environments with similar resource bases may differ in their choice of and adaptation to strategies for defense and development.

DEFENSE, DEVELOPMENT AND DEPENDENCE

The mutual exclusivity of the bodies of literature that have grown up around the processes of economic and political development, on the one hand, and national defense and international security, on the other, is a perplexing fact, one that does not augur well for the further progress of either 'side'.

In the development literature, this can be partially explained by an evident anti-defense bias on the part of theorists, a majority of whom have tended to grant little economic importance to the defense sector in the development process (except frequently and rather rhetorically to classify it simply as a residual 'diseconomy') and who have tended to limit their analyses to the sociopolitical implications of military leadership. Despite their obvious importance for other areas of development policy, military acquisition and production programs have been largely overlooked.

The literature on industrialization patterns and processes is extensive and eclectic. It is singularly devoid, however, of comprehensive analyses of the means by which industrial resources in developing economies come to be devoted to production for defense, a rising trend in more than thirty developing countries. And it is only very recently that even selective attention has been paid to the interdependencies between defense and development variables in either the development or international security fields.

Defense theorists, for their part, have begun only recently to examine the internal economic and political conditions of countries as factors affecting security. Concerned in the past far more with the geopolitical implications of arms and technology transfers to the third world, defense analysis has tended to obscure the impact on the achievement of international security objectives of economic performance, political stability and social welfare in recipient states.

The inattention in the literature to linkages between defense and development is replicated in form and substance in the US policy-making apparatus. Given relatively little formal coordination among the elements of the bureaucracy that administer economic and military transfer programs (or monitor military and civilian technology transactions), there is evidence to suggest that the United States is not fully utilizing its potential to formulate policy based on adequate empirical foundations. As regional security arrangements and national capabilities continue to change in East Asia, altering the context within which relations between the United States, Taiwan and South Korea were first defined, it is particularly appropriate that policy-makers develop a more informed understanding of the ways in which alternative arms and technology-transfer policies have affected the security perceptions and national plans of these key recipient states, and the extent to which the objectives of the United States and its partners are consonant or contradictory.

The stubborn refusal of development and security theorists to combine the unique insights of each discipline is surprising, since at the most basic level of political economy the processes of national defense and development are inextricable: the former exists as a condition for the latter. This is true not only because secure borders provide the modicum of stability necessary for economic endeavor, but also because military production can, under certain circumstances, contribute to the goal of overall industrial development, providing important spin-offs to civilian industries. The extent to which this has occurred in Taiwan and South Korea forms one of the central questions investigated in the succeeding chapters.

Whether or not spin-offs are present to compensate for the 'diseconomy' of defending the integrity of the state, it is important to recognize that for industrialized and industrializing nations alike, the achievement of a measure of self-sufficiency in defense capabilities is a basic corollary of the principle of national self-determination. After the Second World War, for example, the United States encouraged the reconsolidation of the major West European economies by providing military technology and other assistance for the revival of modern defense industries in the allied states. By the 1960s, technology-sharing arrangements were built into nearly all US–European defense transactions, reflecting the will of European states to reject continued reliance on imports from the United States in favor of various forms of coproduction or independent production in the armaments industry so as to keep the indigenous military-industrial sector strong.

The United States has granted coproduction rights to certain developing countries in recent years for reasons similar to those which obtained in Western Europe in the 1950s and 1960s: to create compatibility between recipient forces and US equipment in order to support deployed US forces and to strengthen combined military operations, as well as to encourage multinational acceptance of common strategic and tactical concepts.

At the same time, the commercial and economic incentives for the spread of defense production capabilities have been significant. As recipient countries have achieved higher levels of modern economic growth and gained in political and economic importance via-à-vis suppliers, their ability both to demand and to accommodate a greater share of the weapons manufacturing process has increased as well. With US industry often serving as primary catalyst – if only in recognition that markets could be lost in the absence of some reciprocity – coproduction agreements in recent years have become a more common feature of defense transactions outside the NATO context, as we shall see in the next chapters.

Despite the similarities between the industrial and industrializing worlds in the origin and evolution of indigenous and cooperative defense production, a significant majority of development analysts discuss the military-industrial growth of developing countries in what often seems to be a tone of startled discovery. There is the suggestion – tacit or explicit depending on the ideological posture of the analyst – that investing industrial resources for the development of more independent defense production capacity is somehow a privilege of the industrial world. Given the acute scarcity of industrial resources in most developing countries, it is argued, defense industrialization is an aberrant pursuit, proper only when a certain unspecified level of overall development has been achieved. Even in its most moderate forms, one gets the sense that most observers do not take the defense-industrialization aspirations of third world countries too seriously. With defense modernization possible through imports, it is argued, efforts to achieve independent production lines for weapons must be motivated by transitory concerns such as 'national prestige' and cannot be terribly well fated.

Nevertheless, the achievement of self-sufficiency in defense, in part through indigenous military production, is an imperative that will not yield to calculations of relative inefficiencies and diseconomies, nor can the defense-industrialization path pursued by third world governments be expected to follow strictly the curves of comparative advantage.[1]

This is not to suggest an historical inevitability to the process of defense industrialization, or, what might derive from such an assumption, a form of apologia for the existing policies governing defense investment in the countries under consideration. Rather, it is to suggest that the existence of these industries and their continued growth is no more an historical anomaly in developing nations than it was in countries that achieved industrialization at an earlier time.

The question posed in this study is not the normative one – whether developing countries *should* invest their scarce resources in production for defense; we must expect that under certain circumstances such investment will occur – but rather a more practical one: *how* can such investment best be coordinated with other national objectives such that the twin goals of security and economic, social and political progress are achieved.

Throughout, it must be recognized that because defense production tends to be more costly and less efficient than many forms of civilian industrial investment (rarely conforming to a country's comparative advantage), it is ultimately for military reasons – for the necessary defense of the country – that this burden is borne. As such, any judgments made about the relative success of countries' defense industries must have, as their first criterion, the military benefits that will derive from these industries. Without this objective in mind, it would be easy to conclude that military production is not the best means possible to advance broader development goals, and that without a prior military purpose the economic, and perhaps social and political, effects engendered by defense investment would have at best only low marginal utility at a stupendously high opportunity cost.

INDIGENOUS DEFENSE PRODUCTION AND MODERNIZATION

An independent national defense may always command a high priority in the minds of any nation's leaders, but the 'defense imperative' places the leaders of third world states in a particularly precarious position. While excessive dependency on any single patron state, or any cohesive group of patrons, is risky business in any sector of endeavor, it can, in the area of national defense, compromise the state's status as an independent actor in a particularly visible way. The problems of balancing the goals of development and autonomy are therefore brought sharply into focus.

Third world leaders may seek to achieve a measure of independence in defense by diversifying, where possible, their sources of supply. Another means, one more compatible with the standardization of materiel, the maintenance of national pride, and, possibly, economic modernization, is indigenous production. This is the path Taiwan and South Korea have attempted to follow.

Can military production contribute to economic growth? Better, can it help a nation achieve that delicate balance of autarky and integration within the world economy that even the superpowers and the industrialized West must accept? Can it help create conditions in which the exercise of a nation's comparative advantage is not a limiting and impoverishing process of overspecialization or a simple exploitation of cheap labor resources? In short, can indigenous defense production lead to self-sustaining, balanced growth in conditions of security and stability?

Taiwan and South Korea have both made impressive strides in creating an industrial infrastructure based on capital- and technology-intensive industries useful for defense production. The extent to which the military insecurity of both nations influenced the choice of a development strategy focused on rapid growth of the industrial sector will become apparent in subsequent chapters.

Unfortunately, the theoretical literature on industrialization does not address the question of defense production *per se*, but one can draw interesting parallels from theories of the industrialization process.[2] It is obviously not possible to do justice here to all schools of thought on this subject. However, the major strands of theory as they bear on the study can be presented in abbreviated form.

Economists are divided over the merits of rapid industrialization as a cure-all for underdevelopment. Advocates of the so-called 'trickle-down' or 'unbalanced growth' theory have suggested that the development of a leading industrial sector based on export trade can serve as a vanguard for the rest of the economy and diffuse economic benefits throughout society.[3] After achieving a certain stage of modernization in the leading sector, the economy can 'take off' into self-generating growth where benefits are in turn distributed more widely among the more traditional sectors.[4]

Critics of this view have pointed to deleterious effects of developing one sector of the economy at the expense of others. Doing so is said to lead to 'dualistic' economic structures in which the wealthy strata of society continue to augment their wealth without concomitant benefits 'trickling down' to other strata. The leading sector, in turn, uses the national wealth to buy expensive imports that generally do little to encourage further economic growth. The gap between the modern and traditional sectors widens.[5]

A more recent body of industrialization theory of particular relevance to defense industrialization centers on the economic 'linkages' produced by foreign industrial investment. By developing ancillary domestic industries to supply the foreign enterprise with growing quantities of components of increasing complexity, the country achieves advances in both technological capabilities and in access to foreign exchange for further investment. Over time, this favors the development of a modern infrastructure able to accommodate increasingly complex and independent forms of manufacturing,[6] or so the theory goes. In fact, the economic ripples produced by foreign investment may be very limited if the host country is used simply as a workshop in which manufactured components are inexpensively assembled for re-export. Even in the case of co-ventures, a failure on the part of the host government to provide for the indigenous manufacture of needed inputs can cause overall dependence on imports to rise rather than fall. The result can be a 'treadmill effect' in which the country's efforts to achieve independent industrial capacity simply lead to further structural underdevelopment and external dependence.

Other theoretical problems bear directly on defense industries, which are, on the whole, capital-intensive. Since industrial efficiency in capital-intensive production depends on economies of scale and emerges only at the end of a long learning curve, a substantial amount of time is likely to pass before such an industry becomes economical. In the interim, economic returns on investment will be non-existent. In a small country, moreover, with a scarcity of both skills and capital, it is likely that most heavy industry will be excessive in relation to the country's market size, private funding resources and overall skilled manpower base.

Government involvement is quite often required for the establishment of heavy industry under such conditions, and is particularly likely to occur in the defense sphere, where it may easily appear to be indispensable to the national interest. The costs of such involvement can be great. Illustratively, government subsidization of industries can aggravate inflation by increasing the money supply without raising productivity. Subsidization not only taxes the country's public resources but can inhibit the competitiveness of the subsidized industries by creating overreliance on protection. Wage policies that favor the industrial sector can cause labor migration from other sectors, limiting the productive potential of those sectors and leading to excessive urbanization and unemployment if the migrating labor cannot be absorbed. Finally, inflation reduces the competitiveness of the country's exports and thus can curtail revenues and foreign exchange receipts.

Fortunately, not all returns on intensive investment in heavy industrial production need be as slow to arrive as those that are typically analysed at the level of the firm. Infrastructure such as roads, communication systems and service facilities created to serve the needs of high-priority industries will, if properly planned, prove useful for other purposes as well. Any kind of infrastructure developed in areas where it was previously lacking or inadequate can help to transform social and economic institutions for the better. Roads and highways built to accommodate a new firm, but also useful to the local population, can hasten the overall mobility of the society and thus its integration. In predominantly rural regions in particular, the ability to transport goods can ease the integration of a rural economy into the national money economy. Dual-use communication systems such as radio links and telephones developed initially to support an industrial concern can enhance this process.[7]

The extent to which investment for defense production provides benefits to the economy in the form of material spin-offs depends largely on the location of the relevant industries. If the investment is spread through the country in areas which have the greatest need for new allocations, the positive effects are likely to be significant. If, on the other hand, firms are permitted to locate in already developed urban environments, this can create excessive demand on the existing infrastructure and on civilian services, and result in bottlenecks and economic dislocations. The concentration of national resources in an urban environment can provide further incentives for migration to the urban center, resulting in unemployment and overcrowding, as well as constriction of the rural sector. When industry is set up as a nucleus in a populous environment, moreover, there is an inherent danger of pollution and other social hazards related to unbalanced industrial growth.

Returns on industrial investment may also come in the form of improved human resources. To the extent parts of the population are exposed during the industrialization process to modern technology, production practices and financial institutions, the society may profit, particularly if these serve to bring together segments of the population (military, commercial, professional and academic) whose functions previously required little mutual contact.

As with any new industry, indigenous military production will offer employment to workers previously unexposed to modern methods of production, organization and administration. If work in the defense industries can impart skills useful to the civilian economy, overall development will be advanced. However, the results of studies on the character of defense industries conducted in industrialized countries suggest limits to their utility as a training ground. For one thing, defense industries generally require less labor than do comparable commercial enterprises, while the specificity of skills imparted in the defense sector, particularly in high-technology production, results in a degree of specialization that may not be of general utility in other sectors. Defense production stresses the need for precision and innovation in output more than does commercial production, where cost is the primary consideration. Studies of defense employees in the United States attempting to design products for the commercial market indicate a tendency to 'overengineer' products, rendering them too complex and costly to be competitive.

The inability of other sectors to use defense employees' skills to full capacity may result in underemployment or provide incentives for skilled defense labor to leave the country. Meanwhile, the country's expenditures for education and manpower training, if skewed toward encouraging skills for technology-intensive, defense-related production work, may be undertaken at the expense of other critical skills in short supply, such as modern agricultural methods or energy development.[8]

On the other hand, the enlistment of the country's educational institutions in a combined military and industrial endeavor can be a powerful means of unifying the goals of a country within a cohesive vanguard for development, whose potential for mobilization in countries where alternative political organizations are weak or non-existent has been suggested by Samuel Huntington and others.[9]

An economic sector that produces the bulk of its output for the government, at government specifications and with government financing, obviously will forge strong institutional bonds with the agencies of government responsible for planning its output. Segments of the industry, as a result, may enjoy interactions with government élites that would not otherwise exist. In a country where military objectives are paramount, moreover, the intersection of interests between the military-industry sector and government may afford the industry influence going well beyond the parameters of industrial concerns and extending to decision-making at critical points in the country's overall economic and political planning.

The military-industrial complex can and often does grow into a powerful political-economic interest group with a momentum of its own, with influential supporters in key policy-making bodies. The extent to which this is positive or negative for overall political and economic development depends on whether the élite is disposed to use its power to advance the collective interest of the country or simply its own position and influence. Unfortunately, a powerful military tends to coincide in many societies with a strong domestic apparatus for controlling internal dissent.[10] In a high-threat environment such as that of South Korea and Taiwan, where subversion by hostile neighbors is a genuine possibility, this tendency is particularly difficult to fight, as constant surveillance of the population by official agencies is justified on the basis of 'national security'.

The question remains whether defense production as a whole is more or less useful than other industries as a conduit for technological modernization and a source of economic, infrastructural, social and political spin-offs. On the one hand, because some minimum of expenditure will necessarily be made for national defense, the question of the opportunity cost of that expenditure in terms of civilian investment is, to a degree, academic. Any stimulating effects of the necessary minimum of expenditure will be a net plus, and it seems reasonable that indigenous production of some portion of national defense needs will create more far-reaching linkages than will the direct importation of finished equipment and advisors. On the other hand, the effort to achieve any meaningful degree of autonomy in modern warfare is a very elusive goal for a small nation, one that can be pursued only so far. These realities may limit the contribution of defense production to overall economic modernization.

Defense technology is subject to constant design changes and improvements that alter the utility of equipment rapidly. Keeping pace industrially with even modest improvements in defense articles is likely, under the best of circumstances, to require fairly long-term dependence on imports of key components that are beyond the country's current manufacturing capabilities. Thus, while in some types of production (small munitions, for instance) the country might make significant independent progress, in more complex types of production (such as combat aircraft or missiles) reliance on imported technological inputs may escalate, so that the country does no more than shift the content of its dependence from imports of finished technologies to imported components, while structurally making only marginal advances.[11]

Continued external reliance on suppliers for high-cost technological components may have an accompanying problem if, as is likely, the country feels it necessary to continue to import advanced weapon systems to keep pace with adversaries while waiting for its own industries to become more capable. From the standpoint of national resources alone, this creates a very heavy defense burden that gives way ineluctably to one of two possible outcomes. The country may curtail parts of its production program to finance force modernization based on imports. Alternatively, it may begin to curtail its weapon imports to permit production programs. Both possibilities have important implications for the goals of self-reliance and defense readiness.[12]

Transfers of defense-related technologies are subject to the same limitations on use of the transferred industrial property as obtain in ordinary commercial and industrial transactions. Restrictions on the types of technology actually offered or on the re-export of goods produced under license or through a coproduction agreement are salient examples, familiar from the literature on the investment practices of multinational corporations, but made even more vexing to the host country when translated into the sensitive area of national security.

INDIGENOUS DEFENSE PRODUCTION AND NATIONAL SECURITY

Although inextricably linked with development objectives, the primary impulse behind the series of choices that have created the indigenous defense sectors in Taiwan and South Korea has been the need to protect the nation from the pressures, manipulations and infidelities of its beneficent but overweening patron, the United States, and to widen the range of political choices available to the nations' leaders. This is understandable. The urge toward autonomy displayed by the leaders of these two nations is universal. It has taken powerful form in the universally accepted concept of self-determination, the vital spirit of the large 'non-aligned movement' and something to which all nations pay lip service if often nothing more. It is not our intention to question the value of this elusive and difficult notion, but rather to point out the pitfalls and traps in its attainment.

If in building an indigenous military production capability the Taiwanese and South Koreans have simply changed the color of their dependency on the United States – moving from total reliance on imported weapons to a more varied dependence on finished goods, components and know-how – the fact remains that it is the urge toward autonomy that lies at the root of the decisions to create that capability. And, however Promethean, the effort has not been altogether unsuccessful. The ability of Taiwan and South Korea to produce armaments independently has allowed these states to cope more competently with the vicissitudes of US policy, to display at least to a degree the sort of independence necessary for any régime to retain its legitimacy over the long term, and to take steps they perceive to be in their national interest, whether or not these suit the tastes of policy-makers in Washington. Weapons produced at home have served most obviously to arm troops while creating jobs in the domestic economy, but their importance goes further. They have, to a limited degree, generated export revenue, despite the pervasive circumscriptions of US technology-licensing arrangements. In the case of coproduction schemes, powerful commercial interests in the United States have become advocates of a continued policy of national cooperation with the client state. (As the 'sunk costs' of coproduction grow, the bargaining power of the host country over the supplier grows in a natural progression from abject dependence to considerable leverage, depending upon the importance of the investment in commercial and strategic terms.)[13] Lastly, a credible arms production capability of a certain degree of sophistication can be used both to threaten the design and production of weapons distasteful to the United States,[14] so as to receive concessions for subsequently refraining from such production, and, when the threats cease to work as they should, actually to produce the weapons in question, for use or for export.[15]

Of course, free indulgence by Taiwan and South Korea of their 'semi-autonomy' in defense is not possible. These countries continue to depend on the United States for political patronage (however ambivalent this is in the case of Taiwan), concessional aid (in the case of South Korea) and advanced weapon systems and the technicians to maintain them. Indeed, the visible signs of US cooperation (for example, advisors and investors) can serve, under the right circumstances, as a stabilizing force, demonstrating the commitment and concern of the United States and its support for the régime.[16] The thirst of third world leaders for such manifestations of continuing US patronage stands in clear tension against the simultaneous need to demonstrate their independence, to show they are the puppets of no other power.

In Taiwan and South Korea, changes in US policy suggesting a lessening of commitment – troop withdrawals, refusal to transfer particular military items, and normalization of relations with the People's Republic of China – have typically generated fears of opportunistic subversion by enemies of the régime. These fears have, of course, been unambiguously communicated to the US government. Over time, however, a prudent régime, tiring of the game, must begin to build its legitimacy on firmer foundations than that of US patronage.

Up to now we have skirted the issue of whether indigenously produced arms can actually improve a nation's fighting ability. We have seen that they can be used as status symbols and bargaining chips, but can they, in any real sense, be used in the country's defense? This depends on what is produced. Analysts of different political stripes have argued that the nations of the third world have been ill-served by their dependency on the United States for military assistance. The legacy of dependence, they argue, has been local forces burdened with inappropriate defense tactics based on ineffectual and high-cost technology unresponsive to actual defense needs. Inspired by anticipated alterations in US military assistance policy stemming from the Nixon Doctrine (1969), an important Rand Corporation study concluded that US security assistance in the 1970s should aim at fostering self-reliance among developing countries.

A self-reliant nation is one possessing a national will to depend as little as possible on external assistance in matters of national defense and internal security. Its government will make a realistic effort to mobilize the population to defend itself, utilizing a variety of means ranging from local militias to standing reserves of the regular army. It will adopt a corresponding military doctrine and defense strategy. Doctrines of self-reliance developed in Indonesia and Yugoslavia are relevant to other parts of the third world. In both these countries, the military establishments have refined concepts of defensive warfare using strong deterrent forces that rely on a large number of lightly-armed combatants.[17]

The United States, the Rand Corporation study suggested, could encourage self-reliance by emphasizing transfers of military equipment that was inexpensive, easily maintained, independent of US logistical support and, over time, appropriate for local production (which was to be encouraged).[18]

Unfortunately, even if the United States were disposed to depart from business as usual and embrace such a plan, it is not immediately evident that third world nations could be convinced to abandon their notions of modern warfare and settle for 'appropriate technology'. Bitter experience with the rising cost of technologically sophisticated weaponry may be required to lead developing states toward a better appreciation of their particular comparative advantage in weapons production. While tactics and strategy will reflect the tactician's appraisal of the threat his country faces, they must also take into account the means available to counter that threat. In prosperous times, means can be imported; in leaner times, tactics must often be shaped to make the best use of what can be produced at home.

Reliance on indigenous production does have the advantage, from the point of view of third world military planners, of permitting a choice of both defensive and offensive weapons. Presently, the production programs being pursued by Taiwan and South Korea under the aegis of the United States are essentially defensive (deterrent-oriented) and do not deviate in any significant sense from the force-planning objectives which obtained when the countries were importing technology. As such, the addition of production capabilities should not pose any greater threat to regional antagonists than was posed previously by these countries' forces. If, however, the subjects of our study were to decide to concentrate resources on developing new, offensive technologies thus far not permitted for transfer or coproduction by the United States, this could raise regional tensions. It could, in turn, cause serious bilateral problems with the United States.

Ballistic missiles come immediately to mind. Both Taiwan and South

Korea have, at one time or another, pledged to develop long-range surface-to-surface missiles capable of reaching key targets in the territory of their opponents.[19] The destabilizing effects of such a development could be manifested in two ways; the country may be emboldened to use such weapons in times of crisis; or it may brandish the weapon diplomatically and invite preemptive attack on its production base or other targets.

For Taiwan and South Korea, almost uniquely, the United States has encouraged independent production capabilities as a matter of deliberate policy, derived in part from a desire to extricate its own forces and lower the level of its commitment to the defense of the United States. In so doing, the United States has presumed that it could maintain control over the disposition of defense resources and the nature of both production and deployment. It has further assumed that these programs would continue to be used to strengthen the countries' deterrent capabilities and not to pursue solutions aggressively.

In Taiwan and South Korea, these independent production capabilities represent at least the potential for gaining greater latitude in force structures and deployment decisions, and for lessening the direct leverage of the United States in such matters. If schisms arising from apparently dissonant objectives in supplier and recipient countries are to be avoided, therefore, US policy is going to have to be far more explicit and coherent in the future.

A NOTE ON METHODOLOGY

In 1973, Emile Benoit published a study entitled *Defense and Economic Growth in Developing Countries*, an empirical comparison of defense programs in forty-four developing countries. He set forth the then highly controversial argument that defense programs in the majority of the countries studied had had some beneficial effects on economic growth, which was measured principally by means of changes in the gross domestic product (GDP) and certain investment indicators. No attempt was made to measure the distribution of economic benefits.[20] While it had generally been assumed by most economists that defense represented a drain on resources and retarded development, Benoit's analysis of key development indicators concluded that there were important linkages among defense and development projects that could stimulate economic activity.[21] His analysis served to illuminate the role of the military sector in developing countries, as well as to provide some insight into the process of modernization itself.

The study is based on macrostatistical analysis of aggregated, static indicators, and tells us little about the effect of defense spending, particularly for indigenous production, on political stability.[22] Stability is a curiously misleading word, conjuring up images of an ideal, steady state to which political actors must be encouraged to adhere. In reality, stability – or its opposite – is a by-product of a continuous series of choices by many actors and at many levels. Our analysis deals with choices made at the national level (for example, whether or not to seek to produce certain military equipment). Over time, these choices weigh into the shifting balance of stability at the subnational, national, regional and global levels. This process is hidden by macrostatistical approaches. For example, because the total output of all third world defense industries is only a slight fraction of the total global trade in arms, it would be easy to conclude that these industries posed no geopolitical problems of any immediate urgency.[23] In isolation, such a conclusion would be quite dangerous.

At the subnational level, Benoit's study does not attempt to measure the dynamic structural effects which defense programs might produce in a society over time. However, his conclusions *are* consistent with the view that successful sustainable development depends in good part on bringing about changes in attitudes that impede economic productivity. These changes include the exposure of key sectors of the population to behavioral patterns consistent with modern industrial production and social organization – examples are adherence to schedules and industrial discipline – and the development of base skills such as literacy and the operation of transport vehicles. Otherwise stated, this view has it that traditionalism is the major impediment to successful development, and that almost any institution that breaks atavistic behavior patterns and encourages adaptation to new authority systems and values will contribute to economic growth.

The delineation of quantifiable relationships among defense and development variables is, of course, useful. But the method of aggregate data analysis upon which Benoit has drawn to make his important contribution cannot be used to investigate the complex relations between economic growth and political and social phenomena, and between defense and development variables, to which his study points.

Critics of the case-study approach have suggested that concentration on one or a few cases tends to emphasize 'idiosyncratic' factors and to obscure the more significant, systemic factors presumed to affect all countries universally. Case studies, states critic Arthur Alexander, 'while frequently elucidating the complex policy environment in which particular countries make weapon decisions, are poorly suited to predict the international effects of policies that apply generally: for example, a ceiling on the value of a supplying country's arms transfers'.[24]

For several reasons, however, the delineation of general, systemic factor for cross-country comparison is not presently a valid approach for understanding defense-industrial production in industrializing states. Firstly, in the absence of a detailed case study, it is impossible to know whether or not countries have enough in common – in those ways that are meaningful – to constitute an appropriate group for comparison.

Secondly, it is not clear that a model developed to examine international effects of policies presumed to be applied generally (by the United States and other weapons suppliers) would reflect empirical reality. Supplier *policies*, such as the arms transfer ceiling referred to above, are an insufficient basis for predicting policy *practices*. Rarely are policies applied consistently and universally in the sphere of conventional arms transfers. While there are some obvious exceptions,[25] policies affecting important recipients are usually applied case by case, regardless of enunciated doctrine. Even an idealized arms-control mechanism such as the quantitative ceiling on transfers had only a marginal effect on international arms sales during the Carter administration. Understanding the significance of policy as an independent variable in and of itself requires an approach sensitive to the contingencies to which policies are subject once set forth.

Thirdly, our study takes issue with the notion that a case-study approach is necessarily idiosyncratic, and cannot yield up useful generalizations applicable to a larger number of cases. To some extent, this notion is part of a continuing controversy in social science that originated with the perceived analytical breakthroughs afforded by modern statistical methods, thought by some to be the definitive replacement for all other methods of research. Developing generalizations on the basis of statistical correlation among a large number of cases is still viewed in some circles as the only respectable endeavor of serious analysts, suggesting that other modes of analysis are somehow less rigorous. The initial enthusiasm for these methods has dimmed, however, with the difficulty of detaching measurable variables from the contexts in which they occur. As Sydney Verba put the matter:

To be comparative, we are told, we must look for generalizations or covering laws that apply to all cases of a particular type. But what are the general laws? Generalizations fade when we look at particular cases. We add intervening variable after intervening variable. Since the cases are few in number, we end up with an explanation tailored to each case. The result begins to sound quite idiographic and configurative. In a sense, we have come full circle . . . As we bring more and more variables back into our analysis in order to arrive at any generalizations that hold up across a series of political systems, we bring back so much that we have a 'unique' case in its configurative whole.[26]

As is best explained by the important work of Alexander George, the historical case approach need not be empty of theory or fail to yield useful generalizations just because aspects of the case appear to be unique. Case studies can contribute to the building of theory if they are structured so as to 'formulate the idiosyncratic aspects of the explanation for each case in terms of general variables. In this way, the "uniqueness" of the explanation is recognized but is described in more general terms, that is, as a particular value of a general variable that is part of a theoretical framework.'[27] So structured, a study can enable one to draw out theory on the basis of 'controlled comparisons' in which each case is subject to a set of standardized, general questions reflecting the research objectives and theoretical focus.

As a last word, it should be noted that correlation is not causality. As in a children's 'connect-the-dots' picture, the proximity of two points does not tell us the direction in which they will ultimately be connected in the finished picture, or even whether they will be directly connected at all. In the most sophisticated of such exercises, the picture cannot be divined by inspecting the constellations of points. Only when the rule of the exercise is followed does a meaningful picture emerge: the rules we follow here are those of logic and inference, contrast and comparison. We do not, for the sake of methodological rigor, attempt to approach our subject with *tabula rasa*, emptying our heads of acquired knowledge and viewing our cases as specimens K and T that we find, lo!, to exhibit many of the same behaviors we witnessed in nation-states in the pre-scientific age.

2 The Role of US Policy in Promoting the Defense Industries of South Korea and Taiwan

The purpose of this chapter is to provide an overview of the major aspects of US policy affecting the development of indigenous defense industries in South Korea and Taiwan. These policies include arms and technology transfers as they have affected the forces, tactics, doctrine and overall nature of the two countries' defense establishments, including the production sector. They also include the more general global and regional policies of the United States that have affected Taiwan's and South Korea's threat and security perceptions and have in turn provided incentives for greater independence in defense production. Two themes stand out.

1. The governments of Taiwan and South Korea have strived to cultivate stable and ever more complex relations with the United States. This effort has, from the outset, been accompanied by an equally fervent desire to achieve greater autonomy and self-reliance, for security reasons and for reasons of national pride. The Nixon Doctrine begun in 1969, and the normalization of relations between the United States and the People's Republic of China in the years following, brought the urge toward self-reliance to the fore and set the broad direction of its expression.

2. As the United States has altered its evaluations of the importance of North-east Asia to its defense and foreign policy objectives, changes have occurred in the manifestations of US commitment. The

anticipation of such changes in Seoul and Taipei has had as profound an effect on defense priorities of the two countries as any actual redeployment of US forces. The nuances of policy play a key role in the states' perception of their security environment, even when the US commitment is not actually in flux.[1]

US policy in North-east Asia has an extensive and complex history, but detailed discussion of that history is beyond the scope of this study. Our goal is rather to delineate general aspects of the evolution of policy in the region so as to identify key changes that have influenced the defense planning of Taiwan and South Korea in the direction of greater self-reliance. US policy is the central focus of this discussion not only because the United States has been the major (almost sole) supplier of military assistance, but also because of the profound US role in shaping and sustaining these countries' societies, ensuring, in fact, their safety in the face of repeated threats of invasion.

Since the Second World War, the United States has had as its overarching objective in North-east Asia the preservation of political equilibrium in the region through the containment of Soviet and Chinese influence and the deterrence of efforts by these countries or their allies to gain hegemony through the use of military power. During the 1950s and 1960s, this was achieved largely by the commitment of US military forces to Japan, South Korea, the Taiwan Straits and South-east Asia. Since 1950, South Korea and Taiwan have played a major role in American defense planning for these purposes.

The historical legacies of Taiwan and South Korea are distinct, as is the role the two states have been accorded in US strategic planning. However, their historical fates have intermingled at crucial junctures. The United States did not really begin to appreciate the importance of North-east Asia to its own security concerns until the outbreak of the Korean War in 1950. Although containment was well entrenched as the guiding principle of the American global strategy prior to 1950, this was a largely Eurocentric notion that included the Middle East as its most distant concern but did not really apply to Asia. In spite of the Soviet Union's intransigence in 1945 over the effort to negotiate a policy for a unified Korea (resulting in the division of the country at the 38th parallel), and later over efforts to hold UN-supervised elections for a unified government, the United States still did not seem to grasp fully the significance of the Korean question – or, by inference, any other Asian question – as an aspect of its growing confrontation with the Soviet Union.[2]

The outbreak of the Korean War in June 1950 pushed the United States into a belated recognition of the limitations of its East Asian policy. Previously rather passive on the question of the communist takeover of China, which included an apparent policy of non-intervention in the event of an invasion of Taiwan (for which there was active preparation in China in 1950),[3] the United States was galvanized into a broader commitment to the defense of Asian countries following the North Korean aggression against the South. Although its move to defend South Korea was predictable, less expected were the decisions to reintervene in the Chinese civil war by deploying the Seventh Fleet in the Taiwan Straits and reopening for American and international consideration the legal status of Taiwan – a direct challenge to China's claims.[4]

A growing recognition that events in East Asia formed a critical part of the overall global challenge of communism informed US policy throughout the period of the war and laid the foundations for the active containment policies pursued in Asia in the 1950s and 1960s.[5] The latter were reflected in the creation of a network of anti-communist military alliances (including Korea and Taiwan) and in the provision of large-scale military and economic assistance to Asian countries, both to strengthen defense capabilities and to prevent any possible outbreak of insurrection or subversion. It is this period of American largesse with military assistance and security commitments that had the most profound influence on the development of Taiwan and South Korea as modern military powers. This period lasted roughly fifteen years, from 1954 to 1969. The dependency of the two countries on US assistance – economically, politically, militarily – afforded the United States extraordinary latitude in designing policies consistent with its interests but not excessively encumbered by the concerns of local governments.[6]

The internal development of recipient countries – in Asia and elsewhere – was more an adjunct to policy than an overarching objective. Admittedly, domestic economic pressures during the Eisenhower Administration had inspired the notion of using military assistance to strengthen local forces, lowering the pressure to augment US troop deployments. Military Assistance Advisory Groups (MAAGs) were used in countries like South Korea and Taiwan to provide training and familiarization with the US military organization, tactics and weapons technology. The notion of developing independent capabilities for those countries, however, was not yet central to US defense planning.[7]

One cannot overestimate the indelible impact made upon these societies by the US military and economic assistance programs during this period. In the economic sphere, whether or not internal development was a paramount concern of the United States, US support for economic policies that encouraged investment and production in areas compatible with US import demands led to the adoption of 'export-led' development strategies that achieved extraordinary success. American prosperity during this period was a principal catalyst for the two countries' economic growth. The scale of direct US contributions to Taiwan's and South Korea's development, moreover, was vast. It is reflected in quantitative indicators showing a $3.7 billion investment in Taiwan between 1951 and 1965 and a $4 billion investment in South Korea in roughly the same period, and in the programmatic complexity of the US missions, ranging from direct military subsidies, general economic assistance, widespread training and education programs, to numerous types of investment in support services and infrastructures.[8]

The involvement of the United States, and the American military in particular, in key aspects of Taiwan's and South Korea's development had a profound influence on the nature of their development policies in another important way. The technical and managerial orientation of the two countries' élites is an identifiable by-product of early and extensive exposure to US advisory teams. In South Korea especially, early development of sectors of particular utility to military preparedness derives from the influence of American advisors on the Park régime. South Korea's political development also reflects the experience of American assistance during this period. The army that emerged from the Korean War was the most powerful and effective organization in the country, the only one capable of serving as the focus for American programs. The role of the military was less pronounced in Taiwan, which had preserved a stronger civilian institutional base for political modernization.[9]

Throughout the 1950s and into the 1960s, perceptions in the United States of Chinese and Soviet adventurism in East Asia intensified the perceived importance of Taiwan and South Korea to American defense strategy. Several occasions of actual conflict involving American forces permitted the participation of the fledgling armies of Taiwan and South Korea to gain combat experience using American equipment. Although the local forces themselves were equipped rather modestly, for the most part with surplus US equipment of limited technological sophistication, their involvement in conflict alongside US forces gave them greater exposure to techniques of modern warfare. Chinese pressure on islands occupied by the Nationalists in the Taiwan Straits in 1958 was a particularly significant event for Taiwan's military,[10] but nothing compared to the experience of South Korea's military in the Vietnam War: as many as 300,000 South Korean troops received combat experience in South Vietnam between 1965 and 1972.[11]

Until 1969, the level of US arms transfers to South Korean and Taiwanese forces remained quite modest. Military aid predominated as the major avenue of equipment transfers, although sales had also begun to emerge as another instrument of American policy.[12] South Korea had received modest improvements for its defense forces during the Vietnam War, including items such as medium battle tanks and 175 mm self-propelled guns, but its navy had no submarines or fast patrol or missile-equipped boats, while its air force relied primarily on aging US fighter-bombers and transports. Taiwan's military development was particularly slow in the naval area, its greatest vulnerability, although its air force had comparative advantages in equipment, including Nike Hercules and Hawk battalions (placed under the Air Defense Command), and an array of fighter bombers (F-100s), interceptors (F-104s) and fighters (including the F-5 and F-86) of American design.[13] Even prior to 1970, moreover, Taiwan had established the beginnings of a large facility for defense technology, research and development, the Chungshan Institute of Science and Technology (CIST), and was engaged in two coproduction projects with the United States, one for UH-1H helicopters and another that included the M-14 rifle, the M-60 gun, and 7.62 mm ammunition.[14]

This period of American policy toward Taiwan and South Korea was relatively harmonious. In part, this had to do with the dynamics of dependency: since US assistance and the US military presence were keeping these countries afloat, and the strength of domestic political institutions to exert any counter-pressure on the United States was minimal, neither country had the will or the capacity to do more than acquiesce. Of course, there were periods of friction, but for the most part these remained undercurrents. What was perhaps most significant in this period was the evolution of political consciousness at the national level, a consciousness which gradually perceived the limitations and dangers of dependency, and galvanized a political and military élite to alter the balance by establishing ties of concrete *interdependence* with the United States that would deter any sudden change in US commitment. This was achieved in a significant way in the economic sphere, as both states grew to become major trading partners of the United States. In turn, economic capability meant that both states could begin major investment programs to develop independent defense capabilities – to support the United States, of course, but also as a hedge in the event of change in the US will to defend them.

It was during the Nixon administration that Taiwan and South Korea received the greatest incentives for augmenting their independent defense capabilities. Although US–Taiwanese and US–Korean relations had not been without tensions prior to 1969 – mainly over US policy and deployments in Vietnam, specific interpretations of the defense treaties, and termination of US economic assistance in the mid- and late 1960s – the policies of the Nixon administration turned what had been unspoken anxieties about the basic security commitments of the United States into tangible fears.

Much has been written about the Nixon Doctrine, its origins and intents, and its relative deficiencies; the controversy continues today.[15] For the purpose of this analysis, it will suffice to summarize the aspects that affected the defense plans of Taiwan and South Korea and to describe the trends in arms and technology transfers that resulted from this change in policy.

The Nixon Doctrine inaugurated the period of most profound flux in the security perceptions of Taiwan and South Korea, changes which had a decisive influence on the defense planning priorities evident in those countries currently.

It was comprised of three essential concepts:

(a) 'total force planning', implying that the United States would commit 'all available resources' – intended to include those of allies – to deter any conflict that might interfere with US interests. This concept had as sub-elements the notions of 'combined force planning' and 'integration of Free World Forces' which were to guide the pooling of defense resources globally (and de-emphasize the leading role of the US);

(b) 'regionalism', implying the encouragement of regional cooperation and alliances among third world countries, aided by generous security assistance to local forces, to replace what had been direct US participation in regional deterrence; and

(c) 'self-reliance', which was to be achieved by adapting security assistance programs to specific country needs with the ultimate objective of promoting sufficient independence in self-defense capabilities as to eventually do away with the need for a US role in areas other than training and materiel assistance.[16]

As the Nixon Doctrine was implemented during the 1970s, it became increasingly clear that it was riddled with inconsistencies. Troop withdrawals on a significant scale, as envisaged, did take place, largely in Asia, but domestic economic pressures in the United States dictated that military assistance could no longer be concessionary; thus the previously mentioned shift from grants to sales.[17] This presented the region – and Taiwan and South Korea in particular – with the twin shock of not only having to replace US forces with local troops in key areas, but having to do so with greatly reduced resources.[18]

At the same time, there was an inherent conservatism in the policy suggesting a conscious effort not to make changes in the US defense posture that would imply an abrupt break with the past. Although the policy did suggest a revision of the US stance (to take account of the perception of a reduced threat to the United States stemming from the Sino–Soviet split), it continued to emphasize a leading role for US forces, largely consistent with that of previous administrations.[19] Confusion over the policy's intent fueled domestic controversy in the United States, which added to foreign misconceptions of American intentions. Debate over troop withdrawals from South Korea, for instance, continued throughout the Nixon and Ford administrations, accompanied by significant pressure from some congressional and public circles for further reductions in troop deployments.[20] The official effort to mute impressions that the United States was departing from its prior commitment to the security of Asia thus largely failed, at least in part from ambiguity; and the effort to promote self-reliance among allies, an initiative that in principle would certainly have been welcomed by these allies, rang hollow in light of the unwillingness of the United States to continue grant assistance adequate to this task.

There were, however, more acute manifestations of the fundamental changes in American strategic planning during this era, that overshadowed even the difficult problems of troop withdrawals. The period was a watershed for American strategy, one in which virtually every assumption of the American security system was cast into doubt. The undeniable split between the Soviet Union and China – rendered explicit and concrete by the 1969 border clashes – undermined the traditional notion of a monolithic communist threat, with Moscow firmly in command, which had served as an axiom of the American containment strategy. The limitations of containment, moreover, had been rendered all too obvious by the débacle of the Vietnam War, which demonstrated vividly the need for a redefinition of the limits of American power in Asia and in distant areas in general. These developments coincided with the culmination of a decade-long Soviet military investment effort, which for the first time could be seen to be credibly challenging American strategic supremacy. Worse, all this was occurring at a time when the American public and critical elements of the leadership had begun to oppose further assertions of American military power abroad, and, by implication, any augmentation in military spending to support what was thought to be an atavistic notion of American global responsibility.

The combined effects of Soviet military expansion, the overall decline of the American presence in Asia, and the increasing domestic resistance to military spending and overseas involvement led the United States to pursue a policy that utilized rapprochement with the PRC as counterweight to the rising power of the USSR. At the same time, the disturbing trends at the strategic level required equally bold initiatives to meet directly the threat of Soviet power. This was attempted in part by pursuing a policy of détente, a policy designed to enmesh the Soviets in a network of negotiations and agreements sufficient to induce them to join the international order.

Multipolarity in Asia was the first evident result of the US rapprochement with the PRC, complicating significantly the security perceptions of key allies, Taiwan and South Korea in particular. The changes were sudden. Japan began to reconsider its relations with the United States and the PRC as early as 1971, and somewhat reluctantly recognized the PRC by 1972. By 1973, eighty-five nations had done so, compared to fifty-three prior to the Nixon initiative. Only thirty-nine still recognized Taiwan, compared to sixty-eight prior to 1971.[21] In time, after the United States reduced its opposition to UN membership for the PRC, Taiwan lost its seat. Coupled with the US withdrawal from Vietnam, it was abundantly obvious that the recognition of China was the culmination of irrevocable changes in the carefully constructed Pacific security system of the 1950s and 1960s.[22]

Although the Nixon and Ford administrations emphasized common US–PRC interests in opposition to the USSR, and promoted closer contacts between the countries over the period between 1972 and 1976, they did so without significantly altering formal US ties to Taiwan. Nevertheless, Taiwanese leaders had already set in motion internal measures to prepare them at least partially for what they accurately perceived as the inevitable: the incremental erosion of ties to Taiwan in favor of the mainland, as concerned not only the United States, but also other countries whose trade and support were of critical importance to Taiwan's economic and political stability.

Recognition by Taiwan's leaders that they could no longer rely with confidence on the United States for military security came simultaneously with the understanding that the PRC's enhanced position meant more constricted access for Taiwan to other potential military suppliers. The latter, it was correctly assumed, would reckon on trade reprisals from, and larger arms sales to, the PRC. Together, these perceptions led to more pronounced internal investment in independent defense production. Major alterations in Taiwanese resource allocation patterns and managerial structures were initiated in 1970 and 1971, to promote better coordination among elements of the economic and defense sectors, largely to develop an industrial infrastructure able to support weapons production. Beginning at this time, Taiwan developed arsenals capable of supplying the army with a wide range of small arms, ammunition and artillery pieces, and undertook significant investment in solid-state communication devices, armored cars and tanks, and air frames for jets.[23]

In South Korea, the economic and political effects of the US retrenchment in the region – combined with the successful efforts of North Korea to gain a seat in the United Nations – were perhaps even more pronounced, or at least resulted in more drastic measures. It is thought by some, for instance, that the withdrawal of the Seventh Division from South Korea prompted the leadership to institute martial law in order to insure internal stability. The move to centralize authority in the executive and strengthen the power of the presidency, as enshrined in the Yushin Constitution in 1972, may have altered the political climate of South Korea irrevocably; it certainly impeded any further move toward political development along more democratic lines.[24]

The period of the Nixon Doctrine was a watershed, moreover, in terms of changes in South Korea's economic and military development. Although the development of industrial infrastructure related to defense had been initiated as early as 1961, as noted above, the unprecedented acceleration of industrial investment during the 1970s was in part a direct response to fears induced by the new US strategy. In 1973, the South Korean government decided to push further into heavy industrial development of the kind which would sustain a more broad-based defense production effort. Heavy- and chemical-industry products grew to 25 per cent of exports in 1978, up from 16 per cent in 1972.[25] Moreover, the country began to produce or assemble weapons, munitions and equipment for its ground forces, to build small naval craft, and to develop the capacity to produce aerospace materials.

One key element of the Nixon Doctrine that worked to Taiwanese and South Korean advantage was the liberalization of arms and military technology export codes, so that both countries were afforded far greater access to sophisticated US technology than had ever before been the case. It was, in fact, a deliberate objective of US policy to trade liberalized access to US military materiel for acquiescence in the reduction of the US military presence. This was a global policy under the Nixon and Ford administrations, and unprecedented amounts of advanced military equipment were transferred to countries that had previously received only surplus equipment. However, the policy contained a special twist for Taiwan and South Korea. During the early 1970s, Taiwan and South Korea were the only countries outside NATO and Western Europe to be awarded coproduction contracts with the United States for what could be considered significant combat equipment. Korea received contracts for coproduction of the M-16 rifle in 1971 and 7.62 mm ammunition in 1972, while Taiwan began assembly of the US-1H helicopter in 1969 and coassembly of F-5E fighter aircraft in early 1973.[26] During this period there was a growing level of friction in US–Taiwan and US–Korean relations over the types of equipment and technology eligible for transfer and coproduction. Although the Nixon/Ford arms transfer policies appeared quite liberal in some regions of the world, significant restrictions remained in Asia as a result of the US concern for preserving the overall stability of North-east Asia. US policy toward Taiwan was guided by the need to avoid disrupting the process of normalization in US–China relations, upsetting the military ratio of forces then prevailing in the Taiwan Straits, or contributing to Taipei's offensive military capability against the mainland.[27] Similarly, in transfers to Korea, the United States emphasized weapons that were seen as unlikely to prompt hostile reactions from North Korea or to result in destabilizing transfers to North Korea from the Soviet Union. For the most part, this meant deferral of Taiwanese and Korean requests for items such as the F-16 (or F-18), or advanced air-to-air or surface-to-surface missiles (beyond the 180 km range typical of already deployed systems such as the Nike Hercules).

As production capabilities in both countries developed incrementally – generally with the blessing of the US government – major changes were occurring in the structure of US arms transfer policy that would have important implications for future dealings with Taiwan and South Korea. As stated earlier, one consequence of the Nixon Doctrine had been a sharp increase in arms sales worldwide, from a level of about $1 billion in 1971 to more than $13 billion in 1978.[28]

Congressional attention to what many soon came to regard as a promiscuous arms sales policy emerged as early as 1974. The Senate Appropriations Committee proposed more stringent guidelines for arms sales, calling for more careful monitoring of the way in which US military credits were affecting regional arms build-ups.[29] Attention to this issue was galvanized by the Nixon administration's practice of granting the Pahlavi régime in Iran carte blanche in acquiring American materiel.[30] In 1976, Congress passed a comprehensive piece of legislation setting forth clearer criteria for arms transfers, consolidating existing legislation on arms transfers, and increasing congressional oversight of this process.[31]

The Carter administration entered office in 1977, promising to reduce arms sales. Following a comprehensive Executive Branch review of arms transfer policy and practices in 1977, and a three-month moratorium on sales, the Carter administration announced a new arms transfer policy in May 1977. The new policy, which reflected many of the concerns previously expressed in Congress, set forth guidelines for restraining the flow of conventional arms from the United States. It established both quantitative and qualitative criteria by which US sales would be judged, but stressed that this unilateral policy of restraint would be accompanied by efforts to gain multilateral cooperation – from both suppliers and recipients – which would have to succeed if the policy of restraint were to be continued. Emphasizing that the United States would henceforth regard arms transfers as 'an exceptional foreign policy instrument', in which 'the burden of persuasion will be on those who favor a particular sale rather than those who oppose it', the administration set out to reverse the perceived excesses of its predecessors and to invigorate domestic and international support for this new posture.[32]

The political fanfare that accompanied this announcement, and the extraordinary controversy that surrounded the policy throughout the Carter administration, far outstripped the extent to which the controls actually restrained arms sales. Nevertheless, the grandiose objectives associated with the initial pronouncements haunted the policy throughout its term, inviting criticism from partisans on both sides of the issue. That the policy needed to be 'oversold' because of the peculiar nature of American domestic politics – there were aspects reminiscent of other exaggerated foreign policy pronouncements, such as détente – did little to allay the concerns of recipients abroad who, for the most part, had not been consulted seriously about how the policy would affect their particular needs.

Because the policy aspired to general principles that were to be applied universally, with only an allusion to the possibility of presidential exceptions granted under 'extraordinary circumstances', only Israel (outside NATO, Japan, Australia and New Zealand, which were exempted from the policy altogether) received special attention. That the president would choose to single out Israel, toward which the United States was to 'remain faithful to treaty obligations and honor . . . historic responsibilities', while failing to mention any of the Asian countries for which treaties and 'historic' responsibilities were at least as binding, was troublesome to these countries. Moreover, the guidelines implied serious restrictions on their future access to US arms and technology: the United States would deny sales if the equipment requested would 'introduce into a region newly developed, advanced weapons systems which would create a new or significantly higher combat capability', while coproduction agreements for 'significant weapons, equipment, and major components' would be subject to a blanket prohibition.[33] In addition, the administration's quantitative restraints, embodied in a ceiling on annual sales, implied that decisions would rely on budgetary data and dollar measurements rather than on assessment of the actual requirements of recipients.

In practice, of course, the policy was applied quite selectively and included a number of important exceptions to the guidelines even in the earliest stages of implementation – notably the approval of the sale of AWACS aircraft to Iran and F-15 fighters and related systems to Saudi Arabia – which strained the political credibility of the administration.[34] The Carter administration did deny or defer a series of requests from both Taiwan and South Korea in the early years of the policy, but even with the benefit of hindsight it is not clear whether arms control concerns really had a decisive impact on these transfer decisions, or whether they were simply incorporated into pre-existing political considerations having to do with the stability of the military balance on the Korean peninsula, and with the wish not to provoke the PRC.

Proposed sales were scrutinized, of course, and resulted in some disapprovals on arms control grounds. For the most part, denials to the two countries during this period involved systems that had already been considered and deferred under the previous administration, items such as the Harpoon, I-Chapparal and Maverick missiles for Taiwan, and advanced surface-to-surface missile technology and coassembly rights for an advanced fighter for Korea. Because both countries had previously sought to purchase the F-16 fighter, moreover, the decision by the Carter administration to defer these sales was also more or less a continuation of previous policy. This is not to suggest that the arms transfer policy did not play a role in these deferrals; rather, it suggests that such strict arms control concerns were subsumed under the same longstanding political-military concerns that had guided policy previously.

From 1977 on, arms transfer decisions evolved largely as a function of two major changes in overall US foreign policy for Asia: the decision, initially announced in March 1977, to withdraw US troops from South Korea, and the pursuit of normalization with the PRC, which culminated in the recognition of the Beijing régime as the sole government of China in December 1978.

These decisions have complex histories involving a series of objectives both stated and unstated that cannot be presented in detail here. It is enough for our purposes to summarize the central events surrounding the decisions and the effects each had, firstly, on the threat perceptions of South Korea and Taiwan and, consequently, the patterns of arm transfers.

The Carter administration entered office with a view towards Asia that has been characterized by many as one of relative antipathy. Although this is probably overstated, there is sufficient evidence available in the record of interagency reviews conducted just prior to, and at the inception of the administration to support the view that the Carter administration was committed to a further diminution of the American stance in Asia.[35] That this policy was subject to vacillations over time, and did not, in the final analysis, result in the extensive alterations envisioned at the outset, did not diminish the impact of these early policy directions on key Asian allies.

The decision to withdraw ground troops from South Korea was based on three considerations.

(a) The military balance in the Korean peninsula suggested that the South Koreans could defend themselves against North Korean aggression without US ground troops, given continued US logistics, air and naval support and selective modernization of local forces.

(b) The United States could deter any instability in the area in spite of withdrawal, in that it would continue to emphasize its defense commitment to South Korea.

(c) Both the Soviet Union and the PRC shared an interest with the United States in preserving stability, and would thus not act to endanger this stability by tilting the military balance in favor of North Korea or permitting North Korea to pursue its aims militarily.

The policy evoked immediate opposition in Congress, among sections of the military (especially the military command in South Korea), and among the public, as well as among key allies in Asia, in Japan in particular.[36] Much of this was due to the abrupt way in which the decision was announced. The constituency for withdrawal upon which the Carter administration had counted had either been exaggerated in its significance or had changed.[37] The period from the announcement in March 1977 to 1978 was one of retrenchment for the Carter administration: consideration of military withdrawals from other Asian countries was suspended, written assurances of the US defense commitment were sent to Korea and Japan, and a pledge was made to begin full-scale modernization of South Korean forces to compensate for the impending loss of American troops. Meanwhile, officially sanctioned publications in Korea in 1978 attacked Korean dependence on the United States in unprecedentedly strident tones, and stressed the need to redouble efforts to strengthen the production base for weapons.[38] Economically and militarily, the Korean leadership was preparing the society for greater defense efforts than ever before, including preparation for absorbing the generous supplies of weapons and technology promised by the United States.

There was a certain irony to the Security Assistance Report on Korea presented to Congress in 1978 by Secretary of State Cyrus Vance. It stressed, in emphatic tones, US support for Korean military modernization, including the upgrading of its defense industries, and outlined the series of weapons sales approved in the aftermath of the troop withdrawal decision. Given the prohibition on coproduction agreements in Presidential Directive 13, statements such as the following clearly indicated the inherent contradictions in the administration's original positions.

> As part of its [Force Improvement Plan], the ROK is proceeding with a vigorous program to expand domestic defense production and to decrease reliance on foreign sources of supply and ultimately to reduce the defense sector's demands on scarce foreign exchange resources. The drive towards greater self-sufficiency is being pursued with the understanding and assurance that the United States will continue to serve as a source for defense supplies and equipment, particularly of major sophisticated weapons systems required to counter North Korea's unremitting efforts to strengthen its offensive capabilities still further.[39]

Other measures included the approval of the sale of about $2 billion in military equipment, supplementary FMS credits for 1980, an agreement to establish a joint US–South Korean military command, increases in training and military education grants, and the transfer of about $800 million in weaponry from US forces to Korean control. Among the specific items approved for transfer in 1978 were TOW anti-tank missiles, Harpoon ship-to-ship missiles, F-4 and F-5 aircraft, and air-to-air missiles. Coproduction ventures initiated or continued included those for air defense weapons, light helicopters, infantry weapons and naval patrol craft. In addition, the US augmented its own air power in Korea by increasing the number of F-4s stationed there.[40]

As the United States made these attempts to reassure the South Koreans – and in so doing strengthened the basis for Korean programs aimed at self-reliance – the American intelligence community was reassessing the military balance in Korea. Leaked in early 1979, the reappraisals cast doubt on at least two of the major assumptions of the original withdrawal decision: that the military balance in the Korean peninsula was favorable to the South, and that the United States could count on China and the USSR to maintain restraint in North Korea. The new indications of strength in North Korea, which included a recognition of far more capable defense production capabilities than were previously assumed, cast doubt on the ability of either China or the Soviet Union to exert decisive control over that country and thus to ensure stability on the peninsula. In light of the considerable domestic pressure in the United States against withdrawal and the absence of other diplomatic options to improve the situation on the Korean peninsula, the Carter administration suspended its plans for troop withdrawals in July 1979.[41]

Given the changes in the withdrawal plan, South Korea did not receive all of the equipment promised; most notably, the $800 million transfer of weapons from US to Korean forces. On the other hand, the country did receive the majority of the new weapons it had requested, with the major exception of the F-16 (see below). The net impact of the withdrawal incident was to give Korea's defense industry a healthy boost.

Over this period, the Carter administration was engaged in extremely delicate and secret negotiations with the PRC to normalize relations. Many have noted the sense of urgency felt by the Carter administration in this matter, an urgency apparently not lost on the Chinese. It may well have given the Chinese extra leverage in the negotiations or, as one analyst described it, made it possible for Beijing to 'hold firm against making any substantial concessions'.[42]

In the defense area at least, this seems quite plausible. The United States emerged from the normalization negotiations with an agreement to terminate all military contracts with Taiwan, to withdraw all troops from Taiwan and to abrogate the 1954 Mutual Defense Treaty. Prior to congressional review, the administration's draft of the Taiwan relations bill included virtually no formal mention of arms sales or any further security cooperation between Taiwan and the United States.[43] Moreover, during the negotiations the Carter administration had agreed to impose a one-year moratorium on sales of military equipment to Taiwan.

The absence of consultations between the Executive Branch and Congress and between US representatives and the leadership in Taiwan, coupled with the sudden nature of the decision, invoked intense reaction. In Taiwan, the Foreign Minister expressed his country's view in this way:

> We strongly oppose [President Carter's] decision which we believe has most seriously impaired the rights and interests of this country. We are convinced that it will also impair the long term interests of the United States and endanger the peace and stability of the Asian–Pacific region. Although President Carter's decision is so far-reaching, we were advised of it only seven hours before it was made public. This is not the way for a leading world power to treat a longstanding ally.[44]

The reaction in Congress was equally virulent, in large measure because of the failure of the administration to involve any members in the deliberations. As a consequence of this failure to consult, Congress drafted legislation both to restrict the Executive Branch's future latitude in negotiating agreements with Asian countries and to recast central parts of the normalization agreement to take specific account of Taiwan's future status, most particularly Taiwan's needs.[45]

In the course of the hearings held on the new agreement, it became clear that key senators were seeking a solution to Taiwan's defense problems by devising a 'functional substitute' for the Mutual Defense Treaty, which was to lapse in December 1979. The result was the Taiwan Relations Act, which asserted that:

(a) peace and stability in the Western Pacific area were in the political, security and economic interests of the United States;

(b) the US decision to establish diplomatic relations with Communist China rested upon the expectation that the future of Taiwan would be determined by peaceful means;

(c) any effort to determine the future of Taiwan other than by peaceful means, including boycotts or embargoes, contributed a threat to the peace and security of the Western Pacific;

(d) the United States would provide arms of a defensive character to Taiwan; and

(e) coercive efforts to jeopardize the security, or the social or economic system of the people of Taiwan would be resisted by the United States.[46]

Internally, Taiwan began to take steps to cope with the new situation almost immediately after Carter's announcement. Of particular relevance to self-reliance was a massive countrywide movement to raise funds to bolster the national defense industry. Within six months, every city in Taiwan had contributed to this effort, and $100 million was raised. As one Taiwanese analyst described it:

> The significance [of the money raised] is in the sum total of contributors who represent a national consensus that self-reliance in technology must be regarded as a top priority item in the nation's programs. In 1981, when Taiwan is facing the specter of both a deficit in its international payments and a deficit in its international trade, the newly elected National Legislature worked hard to curtail government spending. However, no voice has been raised to cut defense spending or the budget for defense related research and development activities.[47]

Thus, even with constricted resources, Taiwan – as a result of the mobilization of its population towards a heightened defense effort – was making major strides towards its defense industrialization objectives. Unlike South Korea, however, it was being forced to do so with much-reduced American assistance and materiel. Of the eighteen defense items requested by Taiwan after the moratorium was lifted in January 1980, only six were approved by the end of the year. Included in the denials or deferrals were again the Harpoon missile, Standard air defense missiles, armored personnel carriers, and an advanced fighter to replace aging F-5s (As and Bs) and F-104s.[48]

In mid-1978, the State and Defense Departments had recommended to the president that Taiwan be allowed to coproduce a new fighter aircraft under the aegis of the Northrop Corporation. The F-5G would have been a follow-on coproduction contract to the existing production of the F-5E, which was scheduled to be completed in 1980. The president disagreed with the recommendation, and instead approved the coproduction of additional F-5Es, permitting this to continue through 1983.[49]

Repeated requests from Taiwan for more advanced aircraft were deferred or ignored until mid-1980, when, in response to a letter signed by seven senators expressing concern over the administration's failure to reach a decision on this matter, the president told the Senate that American contractors would be permitted to discuss with Taiwan sales of more advanced fighters. Also in 1980, the administration announced that a new package of military equipment would be sold to Taiwan, valued at $280 million.[50]

When President Carter left office in 1981, no decision had been taken on the fighter aircraft. The $280 million in arms was still largely pending delivery. The announcement of both decisions – the arms package and the permission to contractors to discuss new aircraft with Taiwan – had by that time provoked no significant response from the PRC.

Mainland sensitivities rose markedly, however, beginning in mid-1980, partly as a result of the rhetoric of Ronald Reagan's presidential campaign, which included condemnations of the Carter policy towards Taiwan and promises to improve US–Taiwan relations.[51] Once in office, the Reagan administration found itself in full-fledged controversy over this issue throughout 1981. By December, after a series of objections to US policy, the Chinese media warned that the United States could no longer sell arms to Taiwan if it wished to continue relations, referring to the proposed sales of arms to Taiwan as a 'severe test' of US – PRC relations. To the Taiwanese, the PRC 'diplomatic offensive' against arms sales seemed to clash with important conditions of the normalization agreement. Nevertheless, the country was not prepared for the subsequent decision to deny the F-5G. Confidence in the United States' commitment to Taiwan had been boosted by the content of campaign pledges and statements made by President Reagan himself after taking office. The decision, which was perceived as permitting the PRC to dictate the nature of US–Taiwanese relations, came as an extraordinary surprise to the Taiwanese.

It is difficult to gauge the current internal effects of this decision and of other aspects of the Reagan administration's arms transfer policy for Taiwan. Taiwanese difficulties are compounded by the fact that China is being offered technologically sophisticated inputs for its defense industry, not only from the United States but by the West as a whole, while Taiwan must abide by the strictures placed on it by the United States, with little, if any, prospects for diversifying its sources of supply.[52] The Taiwanese fear not only that their modernization effort will suffer from US export restrictions, but that they will gradually lose their qualitative edge over China's quantitative strengths before the end of this decade, given Western assistance to the Chinese modernization effort.[53] The immediate effects of the Reagan policy, however, stem from the apparent 'unspoken moratorium' currently being applied in the Executive Branch to Taiwanese arms requests.[54]

The Reagan administration introduced an arms transfer policy in

July 1981 that more or less reversed all of the formal guidelines adopted by its predecessor. Reverting to the case-by-case review of sales, and stressing the need to make clear that the United States was willing to back friends and allies with 'the necessary arms, training and support to deter and, if necessary, counter threats to our mutual interests', the administration sought to disassociate itself entirely from the posture of restraint adopted by the Carter administration. In the critical area of coproduction, however, the administration's policy seems deliberately cautious. The policy states: 'The Administration has determined that coproduction requests will be reviewed carefully on a case-by-case basis. Although the US recognizes that coproduction can provide some economic and industrial benefits for both the United States and other participating countries, it also poses economic as well as policy problems.'[55]

The concerns expressed in this guideline seem to center primarily on the possible compromise of advanced technology stemming from coproduction ventures, and on the possibility of third-country sales arising under agreements in which the coproducer insists on export rights for the product. The guideline therefore appears most relevant to US dealings with NATO countries. The efforts of the previous administration to promote technology-sharing and codevelopment projects in NATO met with considerable resistance from the military services – which are always reluctant to share state-of-the-art technology with *any* country – and from some defense contractors that tend to resist coproduction programs for commercial reasons and on the basis of their potential effects on the US defense industry. The extent to which this will apply to countries such as South Korea, for instance, which for now are not engaged in production ventures using highly advanced technology or whose export demands should not pose significant commercial problems for US contractors, is not clear.

With respect to overall Asian policy, the current posture of the United States seems to have been beneficial to South Korea. A request for F-16 fighters was approved late in 1981. The administration seems committed to maintaining troop levels in the region, and has encouraged key states such as Japan to reinvigorate their defense efforts.

THE CHANGING FACE OF DEPENDENCE

It is a cruel irony that measures adopted to advance the goal of self-reliance can sometimes result in greater dependence on outsiders. For both Korea and Taiwan, the development of domestic weapons production has not so much changed the degree of dependence as it has shifted its nature and content. Whereas both states previously were wholly dependent on imported weapon systems for their defense, they have now developed a parallel dependence on imported technology and other factors to sustain their production sectors. Along with the imports come the visible signs of a foreign presence in the form of businessmen, advisors and consultants, and products. Depending on the internal political climate, the foreign presence may work for or against the legitimacy and stability of the régime. The American presence in, and support of, both countries has guaranteed, to differing degrees, a measure of external stability crucial to internal order. In fact, it has, in the past, contributed directly to internal order by bolstering the legitimacy of the régimes and by adding to their governing capabilities through transfusions of financial and technical assistance.

The important question in the current environment is the extent to which these countries have managed to strike a healthy balance between external dependence and increased autonomy. Taiwan is in a special position. Although political isolation has been thrust upon it formally, Taiwan has become tightly integrated within the world economy through less formal trade and economic links. Korea has maintained a highly complex version of dependence on outsiders for decades, ranging from the near symbiosis between the Korean armed forces and US military advisors to the extensive web of financial assistance tying Korea to several countries and to multinational lending institutions.

In considering Taiwan's position, one cannot overemphasize the role which the development of economic ties to other countries has played in maintaining national integrity through the years of political and diplomatic setbacks. For example, even the most enthusiastic statements on behalf of US–PRC ties continue to cite the need to ensure a measure of national autonomy for Taiwan so as to protect US economic interests. To a significant degree, Taiwan's economic strength has become the major deterrent to even more drastic shifts in the status quo.

The contribution of the defense sector to this economic lever is not insignificant. The development in Taiwan of coproduction programs involving US corporate and military interests helped to assure a level of interdependency among American and Taiwanese defense planners which was helpful in protecting Taiwan's interests in the United States. Even with recognition of the PRC, supporters of Taiwan – not insignificantly, industrial groups who had a stake in Taiwan's defense programs – helped to attenuate efforts in the Executive Branch to abrogate defense ties to Taiwan. Although Taiwan's interests did not prevail in the decision on the sale of F-X aircraft, the compromise decision to approve additional F-5Es for coproduction could not have occurred if coproduction were not already a reality.

At the same time, the sudden removal of the American military presence threatened chaos in Taiwan's political power structure. In the defense planning sector, years of close links between the Taiwanese and American armed forces were suddenly terminated. Given the political sensitivities that prevail in Taiwanese defense planning, this presented an extremely difficult situation. Suddenly, there were no American advisors to turn to in the event of difficulties in planning. This was potentially explosive for an internal planning structure that can instantly discredit individuals for even the slightest evidence of error.

The sense of isolation among Taiwan's military planners plays into the hands of conservatives on Taiwan, reinforcing tendencies to pursue drastic measures in security policy, largely to relieve the frustrations of American abandonment. Without the tempering influence of American advisors, the moderates have fewer allies to promulgate their more realistic notions of credible deterrence for Taiwan. According to observers, one early result of the conservatives' recrudescence in defense planning will be pressures to reinvigorate Taiwan's nuclear programs.[56]

For reasons of protocol and tradition, Taiwanese leaders have difficulty accepting unofficial advice on military modernization. This applies to military experts in the private sector, even when these are retired military men who have had prior experience in Taiwan.[57] Seeking unofficial advice is seen as ratifying the status quo, thus reducing the likelihood of future access to American officials. Additionally, a tradition of long standing has it that the legitimacy and importance of advice is directly proportional to the rank of the official dispensing it. Access to American generals in the planning process had been a fact for too long for this protocol to be suddenly eliminated.

Still, sheer necessity will force Taiwan to accept the presence of private contractors as advisors, both to maintain and improve existing forces and to plan future modernization. Wariness of industrial contacts among Taiwanese officials is profound, but it will be overcome. In the civilian sector, for example, acceptance of industrial contractors has increased significantly in the past few years, reflecting recognition that this is the sole means to secure access to up-to-date information on technological advances.[58] The dependence of Taiwan's defense and defense-industrial sectors on foreign technology to keep pace with the rapid innovations in modern weaponry will require a fairly extensive presence of foreign technical advisors, and even skilled foreign labor, for some time to come.

Taiwan has suffered political upheavals following sudden shifts in American policy because of the intricate dependence of Taiwan's leaders on the United States to sustain their legitimacy. Although this dependence often is exaggerated, there were significant instances of political unrest following the US derecognition. A process of gradual political liberalization had begun in the year prior to the shift in US policy, and was to have culminated in elections in late December of that year.[59] Analysts disagree as to the exact relationship between the end of the *abertura* and derecognition. Although the elections were, in fact, cancelled just before the US decision, additional liberalization measures were announced in the spring of 1979, four months later. By the autumn of that year, however, opposition groups, perhaps emboldened by the changes in the external political environment, staged the most violent riots witnessed in Taiwan since before 1950. Some observers have interpreted this political violence as the direct outcome of the Kuomintang's loss of legitimacy following derecognition and the deliberate manipulation of the party's vulnerability under the human rights policy of the Carter administration.[60]

Continued support of Taiwan by the United States, manifested in a growing level of unofficial contacts to take the place of the prescribed official presence, will be crucial to the island's future political stability. The defense production sector is an important source of positive American influence in this regard. Its continued expansion and success may be able to deter those among Taiwan's leaders who might seek high-risk defense strategies which could prove destabilizing.

Korea's reliance on foreign advisors for defense production replicates the pervasive use of foreign assistance throughout the Korean economy. Although at the lower levels of production – unsophisticated munitions and ammunition – Korea is self-sufficient from a technical standpoint, the effort to produce sophisticated weapons such as aircraft and missiles has increased the country's dependence on outside advisors and technicians.

There have been instances of political unrest relating to the American presence in Korea. In the minds of many Koreans, the United States is seen as a master of machinations and omnipotent in Korean politics. Prominent expressions of this perception were the anti-US press statements and demonstrations that took place after the assassination

of President Park.[61] Evidence of anti-American activity has been rare
in recent years. There was a fire-bombing of the US Cultural Center
in Pusan in March 1982. Both the Korean and American governments
maintain active vigilance in this area, reflecting a certain unease about
the large US presence in Korea.

More fundamentally, there is a certain intrinsic instability in the
US–Korean relationship arising from the ambivalence felt by many
Koreans about the extent of the country's dependence on the United
States. In one sense, this dependence antagonizes nationalist sentiment;
in another, there is a strong feeling that the United States is profoundly
responsible for Korea's security, and as such should be more responsive
to Korean needs. As one analyst described it:

> The South's psychological malaise basically results from an angry
> ambivalence with respect to its dependence on the United States
> and its US-promoted dependence on Japan. On the one hand, the
> US presence is a constant irritant to nationalist sensitivities. On the
> other hand, there is a deep feeling that as one of those responsible
> for dividing the country, Washington has a responsibility to support
> Seoul until the country is reunited on terms favorable to the
> South. . . . [T]he psychological burden has been considerable, and
> it has been tolerable only to the extent that continuance of the status
> quo has seemed assured.[62]

Defense industrialization in Korea also has been a source of friction
in US–Korean relations. The United States has helped Korea to achieve
self-reliance in its defense sector only up to a point – that is, only so
long as developments in Korean production capabilities were kept
within the bounds set by American interests and American policy.
Friction over rights to export coproduced materiel has been a fairly
constant irritant, as we shall see. There is, furthermore, a dissonance
in US–Korean interests concerning Korea's ultimate defense objectives.
The American effort to limit Korea's force modernization to those
steps necessary to preserve the regional military balance has not entirely
satisfied Korean military aspirations, as evidenced by the Korean
missile program described in later chapters.

Korean dependence on foreigners is manifested acutely in the role
Japanese business plays in the Korean economy. Although Japan does
not make direct investments in the Korean defense sector, it has
predominant interests in auxiliary industries which support defense
firms, such as electronics. The pervasive influence of Japan is a source

of potentially serious domestic instability. More than any other country, Japan arouses xenophobic reactions because of the bitter memories of colonial experience, as well as the conspicuous material success of Japanese businessmen. Although it has been possible to stem violent demonstrations of anti-Japanese sentiment – and to justify dependence on Japan as a temporary political and economic expedient – there is danger in the dependence Korea has on Japan.[63] This in turn is exacerbated by the overall financial and technological external dependence of Korea, in which defense industrialization plays a part.

On balance, the dependence of Korea on outsiders strains political stability in the country by challenging the strong nationalist feelings endemic in Korean society. The defense industrial sector has, for now, increased the number of foreigners involved in Korea's development, aggravating this problem. In the long run, however, growth in defense industries may, in fact, result in increased independence for Korea, at least in some areas of defense planning. To the extent that this contributes to feelings of self-confidence and self-determination, it would have salutary effects.

This is a long-term prospect. The fact that Korea continues to purchase the most advanced technologies, some of which (like the F-16) exceed its absorptive capacities, suggests that the need for foreign assistance will endure for some time. For now, Korea's position supports the proposition that the development of a defence sector may actually increase external dependence.

Taiwan and South Korea have adapted to alterations in US policy since the 1950s, reacting several times to efforts by the United States to reduce (and for Taiwan, almost to abrogate) its defense commitments. Each successive stage has encouraged successful efforts to develop independent military forces, based in part on indigenous weapon production capabilities. The vacillations of US policy have undermined confidence in the United States, but have not been without indirect benefit to Taiwan and South Korea, because they have promoted greater commitment to the goal of self-reliance.

As the technological sectors in both countries become more capable, one can expect to see a growing ability to produce weapons indigenously, enabling these countries to achieve greater autonomy. Nevertheless, it remains an explicit and implicit assumption of US policy that the United States will continue to exert significant leverage over these countries through direct transfers and the allocation of resources to their production sectors. The outcome of these contradictory trends will hinge on the design of bilateral policies that exhibit more understanding of the internal realities of Taiwan and South Korea than has been the case in the past.

3 The Defense Industries of Taiwan and South Korea in Detail

Publicly available data about the defense industries of Taiwan and South Korea is limited. The sensitivity of the subject in the two countries, coupled with the relative inattention of Western analysts, has resulted in the absence of substantive descriptions of the countries' defense industries in traditional sources. Nevertheless, it has been possible to develop an independent data base, founded on a rather eclectic series of sources, that provides an overview of the defense production programs underway, a description of the key aspects of their structure and progress, and, wherever possible, quantitative data on production levels and costs.[1]

From a general standpoint, it is useful to keep in mind that the development of Taiwan's and South Korea's defense industries has followed a pattern of industrial development common to a number of other developing countries. Although too superficial to be more than illustrative in nature, this pattern does suggest some guidelines for understanding the deeper defense industrialization process. It can be summarized as a sequence of four essential stages:

(a) initial import of arms from foreign suppliers;
(b) gradual creation of maintenance and overhaul capabilities and related facilities, including the manufacture of spare parts, made possible by the provision of equipment, data, training and supervision from foreign sources;
(c) eventual assembly and production of major weapons under license, to include varying levels of technical participation by the host country;
(d) indigenous design, development and production of systems (see Figure 3.1).[2]

	Arms imports	Maintenance and overhaul capacity	No local production capability, but local assembly of imported subassemblies	Limited license building assembly with some locally made components	Some independent defense building, but important components imported	Local licensed production of less-advanced arms; R&D on improvements and derivatives	Local licensed production for most weapons; limited R&D production for less advanced	Complete independence in research, development and production
Small arms and ammunition								
Trucks, jeeps								
Small patrol boats								
Light armored vehicles								
Larger naval platform								
Heavily armored vehicles								
Helicopters								
Training and transport aircraft								
Aircraft engineers and advanced missiles								

The two variables of defense production, degree of independence of production and sophistication of weapons produced, are shown here in matrix form, with a rough breakdown of weapons and degrees of independence. These categories are by no means exhaustive and are presented for illustrative purposes only.

SOURCE Adapted from Stephen E. Miller, 'Arms and the Third World: The Indigenous Weapons Production Phenomenon', unpublished paper prepared for the Programme for Strategic and International Affairs, Graduated Institute of International Studies, University of Geneva (June 1980) p.7.

FIGURE 3.1 *Matrix of arms production capabilities – degree of independent production*

The progress of South Korea and Taiwan from stage to stage has not proceeded in sequence, and, at this point, elements of all four stages operate simultaneously in both countries. It is likely that this will continue for the indefinite future. Even the most successful development program often requires some types of imported components, even in industrial countries, while the maintenance of effective forces may require outright imports of advanced weapons that the countries cannot or prefer not to produce.

Program complexity varies not only with the technical proficiency of recipients but also with the terms of the licensing agreements suppliers have been willing to provide. Suppliers' restrictions often limit host countries' technical participation in defense production programs.[3] Among the types of licensing agreements are:

(a) provision of data for the production of simple components;
(b) technical assistance to create a production facility for components or entire systems with foreign assistance and supervision;
(c) final assembly and testing of systems, components for which are provided by the supplier; and
(d) manufacture of systems in a host country facility with the percentage of locally built components gradually increasing over time.[4]

This chapter presents information on the defense industries of Taiwan and South Korea. It is intended to serve as the foundation for a subsequent analysis of the economic, political, and military ramifications of these programs.

TAIWAN

Taiwan's defense industries have evolved over the last two decades as a result of direct US encouragement and in response to the diminution both in the American military presence in East Asia and in American military aid. As such, American policy provided both the incentives and the means for Taiwan to develop a defense industrial capacity. Largely to insulate themselves against the vagaries of American policy, Taiwanese defense planners have sought greater self-reliance in defense planning, particularly in the last decade. This was made possible by the provision by the United States on a selective basis of the technological inputs and expertise to initiate and advance indigenous production programs. In the 1960s, for example, American foreign policy affected the Taiwanese defense base directly when the US Air Force utilized Taiwanese facilities to repair aircraft during the Vietnam War, giving Taiwanese engineers experience in the repair and overhaul of US F-4 jet fighters.[5] Moreover, when the US redeployed a large part of the US Seventh Fleet to South-east Asia, leaving by 1969 only one destroyer escort in the Taiwan Patrol Force, the Taiwanese were moved to replace the dwindling US force with an indigenous one, based in part on domestically built naval vessels.[6]

The beginnings of a defense-industrial base in Taiwan were in the years following the termination of American economic assistance in 1965. Prior to that date, Taiwan was more or less completely dependent on the United States for all military acquisition, although in fact it had already developed some interest in expanding its access to other suppliers.[7] As one of the means of adapting to the shift in US aid policy, Taiwan built a number of government arsenals capable of supplying its army with small arms and artillery pieces. These arsenals have served as the basis for expansion of the defense industrial infrastructure, and over the following decade and a half have achieved a production capability for items such as fighter aircraft parts, high-speed patrol ships, armored cars and missiles. Each of these production programs is discussed in detail below.

Development of the defense industrial base was accelerated in the early 1970s, upon the initiation of US–PRC contacts and a precipitate decline in US military aid to Taiwan. The Shanghai Communiqué in 1972 prompted a thorough review among Taiwan's defense planners of the country's dependence on the United States for military hardware. Given political constraints on Taiwan's access to other suppliers and the fact that the island's forces were configured almost exclusively around US military equipment, a strategy of 'self-reliance' necessarily took precedence over efforts to acquire arms from non-US sources.[8]

Among the developments of the late 1960s was an increasing effort on the part of national planners to coordinate economic development and defense priorities so as to create compatible production sectors with minimum inefficiency.

Taiwan's economic and defense planners have typically been sensitive to the trade-offs inherent in allocations of resources to military technology projects at the expense of civilian development objectives. One of Taiwan's major fears is the possibility of a trade war with the PRC, in which low-cost, labor-intensive, consumer products from the PRC (textiles, food products, simple electronics) might capture enough of Taiwan's foreign market to constitute a sort of war of economic attrition on Taiwan. Even in the absence of a direct military threat to Taiwan's security, this kind of economic conflict could undermine economic and political stability in a country so highly dependent on trade for survival. In this sense, Taiwan's economic performance, especially in trade, is as critical a 'defense' variable as is its calculable military potential.

As a result, Taiwan has deliberately skewed its overall industrialization strategy toward technology-intensive industries; in the words of one Taiwanese analyst, to 'stay a few jumps ahead in technology so that [Taiwan] can produce goods and services beyond the reach of her opponent'. In order to remain competitive in international markets in this area, it is critical to have efficient industries capable of producing relatively advanced civilian goods for which there is sufficient demand. Coordinating this with defense objectives requires that, wherever possible, R&D and technological advances be tailored toward products which have the potential for dual use. This requires extremely careful planning, in both a technical and managerial sense, so as to forge links among the industrial, research and defense sectors.[9]

Until the very recent past, Taiwan's defense industries were wholly owned and operated by the government. In the last three to four years, however, the government has moved to increase the share of civilian industry in defense production and to divest itself of some of the more inefficient production programs, particularly in ground-force equipment.

One manifestation of this need for closer coordination has been the official promotion of greater cooperation in military and private technology research and development under the auspices of the Chungshan Institute of Science and Technology, a government-run research facility for defense technology. Established in the mid-1960s with only a few technicians, scientists and engineers trained by the armed forces, by the mid-1970s CIST employed nearly 2000 professionals, some 50 per cent of whom came from universities and private industry.[10] This has made it possible to ensure that defense projects provide maximum spin-offs and benefits to the civilian economy at a minimum cost. Aside from the technical and scientific advances made possible by training science professionals in this institute, moreover, the consolidation of scientific expertise under a single government-sponsored roof has helped to build an appreciation among Taiwan's academic-industrial élite of the nation's defense requirements.[11] In addition to being primarily responsible for advanced research in missiles and rocket technology, it performs R&D on nuclear energy, chemical components, guidance systems, and other electronic systems. (Chungshan produces prototype weapons for production in government arsenals.)[12] In recent years, CIST has commissioned work for universities and expanded research activities in the private sector, forging links to civilian groups such as the Industrial Technology Research Institute and the Atomic Science and Nuclear Engineering Research Institute at the National Tsinghua University.[13]

The 'derecognition' of Taiwan in December 1978 transformed irrevocably the challenges faced by Taiwanese defense planners. In addition to the revocation of the 1954 Mutual Defense Treaty and the phased withdrawal of all US military forces (completed in 1979), Taiwan's access to US military technology was formally restricted to small amounts of equipment and services necessary to maintain existing defense capabilities. Although no shift in Taiwan's defense strategy seems to have occurred, major changes have occurred in the country's planning apparatus and in defense allocations. Illustratively, the defense budget has grown at an annual rate of about 25 per cent since 1977 (inflation may account for about one-third of this increase) and amounted to about $3 billion in fiscal year 1981, including a special supplementary allocation of about $700 million in that year. This represents a 50 per cent increase since 1979, to more than one-half of Taiwan's total national budget.[14] Budgetary resources devoted to weapons production seem also to have increased markedly. An estimated fourteen weapon-related production programs were being financed in 1981 by the National Defense Industry Fund, with allocations thought to be in excess of $140 million.[15]

Authority for military production and production planning is vested in the Ministry of Defense, with operational control delegated to the Combined Service Forces (CSF) and the separate military branches of the armed forces and their affiliated agencies. Co-equal with the respective services, the CSF General Headquarters is responsible for all logistical support to the army, navy and air force, and to Taiwan Garrison (security and intelligence). This responsibility includes procurement, weapons production and materiel management.

Planning and procurement are subject to rigorous scrutiny through the systems-planning procedures the Taiwanese adopted from the US Program Planning and Budget System (PPBS), a highly-organized, technical series of procedures for procurement planning. Central to this is the 'tyranny of the budget'. Each agency and armed service is required to submit a detailed estimate of its import needs for each fiscal year. Service and agency requests are then subject to the kind of protracted, contentious debate common to the United States, the 'head-knocking' required for agreement on realistic force estimates. An unfortunate legacy of the US planning process inherited by the

Taiwanese is the conviction of each of the services that they are solely responsible for the defense of the country. This leads to exaggerated requests and intense political bargaining, in spite of the strict budget criteria to which planners are forced to adhere.

Logistical planning in Taiwan is highly advanced. A central computer in Taiwan is connected to the central military logistics center in the United States. Requests for spare parts and components covered by continuing supply support agreements are fed daily into the computer network. The projection capability of the network estimates equipment needs in a very precise fashion. Handling an estimated 6000 parts a day, the system makes less than 50 mistakes a month.[16]

Despite the ability of Taiwan's decision-makers to mobilize resources and public support for defense production, there are serious technological obstacles to achieving full-scale capabilities in the manufacture of even moderately sophisticated weapon systems. Firstly, the volume produced, given the small demand of Taiwan's armed forces, makes it difficult to achieve efficiency in production. Secondly, technical difficulties make it hard to achieve quality and reliability in production at a level sufficient to make the products useful militarily. And thirdly, politically inspired limits on Taiwan's continued access to weapon components and related equipment and technology, from the US or elsewhere, greatly hamper the defense effort.

The first consideration is largely one of high cost. In the absence of exports – limited by the nature of the products available for export, largely small arms and ammunition, for which there are many suppliers, as well as by political factors – no industry can recoup the costs of investment, far less gain profits, from such small-scale production. As a result, at least until recently, the government has had to undertake production projects on the basis of government-sponsored 'pilot plants', incurring high costs not only in the form of government subsidies but also in the diversion of skilled manpower from potentially more renumerative R&D activity.[17]

The second factor relates to the limitations of Taiwan's existing technical infrastructure. Achieving military quality in production requires precision industries and highly-qualified manpower trained in very specialized skills. Of particular importance, of course, is the condition of the machine tool and machinery industry, which, although thought to have improved considerably in the late 1970s, is still embryonic.[18] This industry in turn depends on the overall industrial capabilities of the country, including the steel, petrochemical and non-ferrous metal industries, as well as electronics and metal fabrication.

Overall, Taiwan's defense industries are still weak in the area of original R&D and design work, which in part stems from the management practices of the military-run research institutes.[19] In recent years, however, an increasing amount of R&D work has been placed under the responsibility of predominantly civilian institutions, such as the National Science Council, which now has the authority to carry out research efforts on behalf of the defense establishment and serves to coordinate efforts among the industrial, university and defense communities.[20] However, the extent to which these organizations are autonomous from government in any way comparable to Western institutions is still quite limited.

As will be illustrated in the ensuing discussion of specific programs, Taiwan's industry is particularly proficient in reverse engineering and the copying of foreign designs. Supplementing this capability is the fact that until quite recently, Taiwanese engineers and scientists had access to a rather full range of training 'within the limits drawn by the United States Government on security and defense related research'.[21] Although this training has been severely curtailed since derecognition, some training of civilian engineers and scientists in areas of potential utility to military research continues in the United States.[22]

The third constraint, that of Taiwan's access to technological components and manufacturing know-how from overseas, is the least mutable concern at present. Although Taiwan continues to receive inputs and spare parts for its industry from the US under existing foreign military sales, and although it has managed to conclude military equipment agreements with Israel and has been conducting technical discussions with France concerning the potential purchase of French nuclear reactors,[23] access to types of inputs the country needs is never assured. Because Taiwan's military industries have been built up around American technology, access to US goods may be the major bottleneck to the continued expansion and improvement of Taiwan's defense industrial base.

The major management deficiencies of Taiwan's military–industrial planning apparatus inhere in the political characteristics of Taiwan, one of which is the schism between the 'old guard' political leaders and emerging professional military leaders. The former continue to stress atavistic notions of the threats facing the island (invasion from the mainland). The latter, while forced to pay obeisance to the older political faction, try gingerly to promote more modern defense concepts. The 'old guard', many of whom occupy élite positions by virtue of their loyalty to the Kuomintang and its notions of Taiwan's destiny, limit the ability of the slowly emerging professional military to plan programs in a rational fashion. This highly complex political picture will be discussed in greater detail in Chapter 4. Suffice to say at this point that a number of the less enlightened defense decisions Taiwan has made in the past – both in acquisition and in production – reflect the triumph of political over military concerns. Moreover, the political constraints upon personnel at the operational level – project managers, for instance – have militated against innovation and positive risk-taking in defense production, slowing the pace of development of Taiwan's military potential.

At the same time, a growing effort to increase participation by private interest groups in defense research and production is likely to enhance the innovative capacities of the Taiwanese defense industrial sector. Joint government–private research institutes, such as the Industrial Technology Research Institute, serve as an important focus for current technology-transfer projects. To attract scientists and technicians from domestic and international sources to Taiwan, more-over, the Taiwanese government has established the Hsinchu Science Industry Park. It is aimed at strengthening the high-technology manpower base of Taiwan, so that the latter may eventually attract, in the words of Minister of State K. T. Li, 'the kind of talent to the area that now exists at the West Coast's "silicon valley" in the San Francisco area'.[24]

As Taiwanese policy moves increasingly toward reliance on private rather than government expertise, the rigidities inherent in the government-run defense institutions should relax considerably.

The ensuing section describes in detail the production programs and problems in each of three sectors of Taiwan's defense industries – aerospace, naval forces and ground forces equipment.

AEROSPACE

Manned Aircraft

Currently in Taiwan, the program requiring the greatest technical proficiency and having the greatest relevance to critical defense requirements is the coassembly–coproduction program for F-5E fighter aircraft. Initiated in 1973, this program is supervised by the Aero Industry Development Center (AIDC) and operated by the Taiwan Air Force. Centered in a massive (an estimated 65,000 square meters) facility at Tai-chung, where the F-5s are produced, the AIDC is responsible for production of all military aircraft. Its other production and/or development programs include production of a limited number of US T-28 trainers (and T53-L-701 turboshaft engines) at a facility in Kang-shan; and design and development of the XC-2 light, twin-engine transport, the XAT-3 twin-engine trainer, and the XA-3 single-seat attack fighter.[25] Largely as a result of these aircraft and helicopter assembly programs, and of the exposure of ROC engineers to aeronautical repair for the US military during the Vietnam War, Taiwan has a most impressive ability to overhaul and repair all major types of aircraft.

The F-5E/F program is the centerpiece of Taiwan's defense industrialization efforts, and is certainly its most significant military project. At the time of writing, Taiwan has fewer than 400 fighters, of which the F-5E/Fs number about 200, with another 48 scheduled to be produced by mid-1983.[26] Compared to the inventory of PRC jet fighters, estimated to be over 4000, Taiwan obviously suffers severe numerical inferiority. From a qualitative standpoint, however, at least until the mid-1970s, the F-5E provided an important edge of superiority. Designed to counter the Mig-21 (designated F-7 in the PRC configuration), the F-5E exhibited greater maneuverability and, when equipped with heat-seeking Sidewinder air-to-air missiles (of a type not developed by the PRC until the mid-1970s), could have decisive advantages in air combat, particularly in light of the superior pilot training of the ROC air force. Subsequently, however, the PRC has developed a version of the Mig-21 (designated the F-8) the performance characteristics of which are not fully known but are assumed to be superior to the F-7 and to outstrip those of the F-5E.[27] Additionally, the PRC has produced and deployed A-5s, enlarged versions of the Mig-19, which are thought to be potentially effective against Taiwanese naval vessels and defense installations on the offshore islands if China gained control of the airspace over the Straits.[28] Coupled with the growing obsolescence of the remainder of Taiwan's aircraft inventory – ninety F-100 A/F aircraft, sixty F-5A/Bs and a small number of F-104s, all of which are expected to be obsolete by 1986 – these air threats to Taiwan have led to requests for high-performance air superiority fighters, which have been denied by the United States.

In the face of the US disinclination to transfer aircraft other than F-5Es, the coproduction program has become more important than ever. Until such time as the Taiwanese gain access to other suppliers for aircraft, an unlikely event, or the US revises its decision, the F-5E program will be Taiwan's only source of modern aircraft. Moreover, from a technological and economic standpoint, the program has been of tremendous importance in generating new skills for Taiwan's labor force, in exposing the ROC air force to new technical capabilities, and in serving as a focus of national pride in the development of an indigenous aircraft industry. In financial terms alone, the ROC has spent an estimated $1.1 billion on this program since 1973; in some years, the F-5 program cost as much as 80 per cent of the country's total foreign military sales credits.[29]

Gains in technical capabilities aside, limitations in the F-5 program augur badly for the ROC becoming significantly more self-sufficient in aircraft production, unless, of course, technical support is forthcoming from outside suppliers. Although the program has moved from simple coassembly in 1973 to increasing participation by the Taiwanese in the fabrication of components (the most advanced being structural assembly of wings, done almost entirely by Taiwanese engineers – see Figures 3.2 and 3.3), the AIDC has no capability in critical areas such

PHASE I	●FINAL ASSEMBLY ●FLIGHT OPERATIONS
PHASE II	●PHASE I PLUS SYSTEMS INSTALLATIONS AND MAJOR ASSY
PHASE III	●PHASE I AND II PLUS STRUCTURAL ASSY AND SUBASSY OF FWD FUS
PHASE IV	●PHASE I, II, AND III PLUS FABRICATION OF FWD FUS AND SELECTED BONDING
PHASE V	●PHASE IV PLUS FABRICATION OF DORSAL COVERS, AILERONS AND T.E. FLAPS
PHASE VI	●PHASE V PLUS WING STRUCTURAL ASSEMBLY AND SKIN FAB

SOURCE Northrop Corporation, 'Fighter Coproduction and Security Assistance', Presentation document on the F-5G program, NB81-95 (May 1981).

FIGURE 3.2   *ROC F-5 coproduction phases*

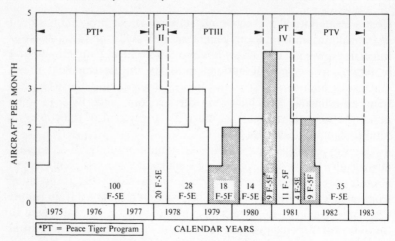

Source   Northrop Corporation, 'Fighter Coproduction and Security Assistance', Presentation document on the F-5G program, NB81-95 (May 1981).

Figure 3.3   *ROC production rate/delivery schedule (248 aircraft)*

as avionics or pneumatics, and only limited capabilities in engine fabrication. Technical difficulties also persist in precision machining, casting and specialty metals.[30] Northrop, the US manufacturer, itself stresses to US audiences that the F-5 technology used in Taiwan is over fifteen years old, that overall Taiwan produces only 10 per cent of the total F-5E, and that no R&D, production design or qualification technology has been transferred as a result of this program.[31]

Among the critical capabilities Taiwan will have to develop in the future to sustain an independent or quasi-independent aircraft industry are those relating to avionics and the manufacture of jet engines. This will be possible only with continued high levels of foreign assistance. Though the indigenous jet trainer and transport programs cited earlier have in fact been successful, they do not provide a sufficient technological base for the design, testing and manufacture of high-performance jet fighters. Moreover, even the training and maintenance facilities Taiwan has been able to develop may rapidly become obsolete if ROC engineers and military personnel are prevented from keeping pace with the rapid changes that characterize advanced Western defense technology.

Missiles

Taiwan's leaders consider access to modern missiles and missile technology to be of central importance to the island's defense. Major modernization priorities include an improved air defense system and the development or acquisition of air-to-air and surface-to-surface missiles of greater sophistication than those in inventory, which include the US AIM-9 Sidewinder, Maverick, Hawk and I-Hawk, Sea Chapparal, and Nike Hercules missile systems. Repeated requests from Taiwan for the more advanced AIM-9L air-to-air missile, Standard air defense missile, and Harpoon anti-ship missiles have been denied. Nevertheless, the persistence of the requests reveals that Taiwanese defense planners consider the kinds of missiles the US has been willing to provide to be numerically and qualitatively inadequate. They have accorded a high priority to indigenous missile development programs. These appear to have been underway for almost a decade.

Unclassified evidence of Taiwan's efforts to develop independently certain types of long-range air-to-air and surface-to-surface missiles over the past decade is not extensive. However, references to such programs do emerge publicly on a periodic basis. A fairly significant confrontation between the governments of Taiwan and the United States occurred in 1976, when it was discovered that fifteen Taiwanese engineering students from the Chungshan Institute were studying advanced inertial guidance technology in a special program at the Massachusetts Institute of Technology. (The students had claimed to be from the National Taiwan University.) The program was cancelled by the State Department, on the grounds that the agreement for 'training to develop complex ideas into marketable products using the specific medium of transfer of inertial navigation and guidance technology' was in actuality more military than commercial in nature and as such violated US munitions export codes. In 1974, an issue of *Overseas Scholars* (Hai Wai Xue Ren, published in Taiwan) published an advertisement from the Chungshan Institute seeking a wide variety of missile-related specialists, with some 77 openings for engineers with such skills.[32]

In one sense, what may be most important about these reputed missile programs is not the military impact of actual production capabilities so much as the political effects reports and rumors of these programs can have. One effect of commencing R&D for systems the production of which the US would not condone is to add to the leverage countries can wield in bargaining for direct transfers of US weapon systems. It has, in fact, been a traditional policy response in the United States to discourage certain countries from developing systems perceived to have a potential for regional destabilization (such as long-range surface-to-surface missiles) by providing sophisticated

US equipment. As such, the strategy of the Republic of China may well be to *develop* rather than *produce* certain items so as to increase their procurement options.[33] This would pertain particularly to items that would be extremely expensive to produce indigenously, such as advanced missiles.

Taiwan presently does have a modest missile production capability. AIDC began licensed production of the AIM-9 Sidewinder missile, for instance, as early as 1974.[34] Most of Taiwan's missile activities are concentrated in the Chungshan Institute, which is reportedly equipped with sophisticated computer technology for use in the design and evaluation of certain types of missile technology.[35] CIST developed and produced the Hsiung Feng anti-ship SSM, which was based on a reproduction of the 'Gabriel' SSM purchased from Israel in 1973. It is thought to be a relatively short-range missile (30–50 km) which may be superior in accuracy and in its ability to evade radar than the comparable 'Styx'-type missile (designated SS-N-1) deployed by the People's Republic of China.[36] Displayed publicly in 1979 during a National Day parade, the Hsiung Feng missile is now used by both the Taiwanese navy and coastal defense forces.[37] Taiwan is also developing two shorter-range missiles, the so-called Green Bee SSM and an anti-tank missile.[38] The anti-tank missile, thought to be designated Kun Wu, is probably vehicle-mounted. One of the ROC's SSM/ASM production programs has produced the Ching-Fong SSM, although its capabilities are not known. These programs represent important steps towards Taiwanese production of high-technology weaponry sufficiently free of US components or designs to contribute to Taiwan's plans for independent defense and export production.

NAVAL PRODUCTION

While Taiwan's defense requirements should focus logically on naval forces, the army still has top priority in Taiwan's defense planning. However, the threat to Taiwan from a mainland naval blockade has gained more prominence in Taiwan's strategy in recent years, and has directed attention to naval force modernization.

Most shipbuilding in Taiwan is directed by the navy and conducted at naval shipyards in Tso-ying, Ma-kung and Kao-Hsiung, with secondary facilities located at Chi-lung.[39] These yards have sophisticated repair and maintenance facilities.

The sophistication of the repair yards can be illustrated by a few examples of modifications . . . the rebuilding of the LST (Landing Ship, tank) force, fitting each of the frigate transports acquired during the 1960s with a second 5-inch turret and magazine aft, and the installation of Gabriel surface-to-surface missiles in several destroyers. Furthermore, the yards cannibalize both stricken ships and others purchased specifically for that purpose. For example, the US destroyer *Warrington* was bought in 1973 after it hit a mine off North Vietnam, and three of the six FRAM II destroyers bought in 1974 (the *Swenson*, *Bole*, and *Lofberg*) are being cannibalized to keep the other FRAM vessels in operation.[40]

In fact, Taiwan's naval yards can repair any vessel in inventory. They are supplemented by the capabilities of the Taiwan Shipbuilding Corporation (in Chi-lung) and the China Shipbuilding Corporation (at Kao-Hsiung).[41]

In the early 1970s, Taiwan built a smaller number of 30-ton patrol boats, the first naval craft to be indigenously produced. These were equipped with radar and carried a single 40 mm cannon.[42] In 1975, Taiwan signed a contract with a US firm, Tacoma Boat, calling for the production of a series of Multimission Patrol Ships (PSMM) under the aegis of China Shipbuilding. These ships have an all-aluminum hull, a technique new to Taiwanese shipbuilding. The program has been beset by problems, however, ranging from escalating costs to poor test performance. The patrol boats are subject to considerable instability in the sea conditions common in the Taiwan Straits, which are both shallow and extremely rough during the monsoon season from October to March. Additionally, the aluminum hulls, possibly because of production problems, have been subject to serious corrosion.[43] Taiwan also has a contract with Westinghouse to produce frigates, although the status of this program is not known. It is thought to be moving very slowly.[44]

For anti-submarine warfare (ASW), the Republic of China has a fleet of destroyers, significant in tonnage but quite old, as well as the US 500 MD helicopter with light anti-submarine warfare capabilities.[45] Given the very real threat posed by Chinese submarines in various circumstances, this is a fairly serious deficiency. Taiwan has ordered the US ASROC RUR 5A surface-to-surface ASW system (delivery was expected early in 1982), but does not produce any such system indigenously.[46] Taiwan particularly needs more sophistication in naval electronics, underwater acoustics and naval armaments as a whole, the

Hsiung Feng missile notwithstanding.[47] A recent agreement with a Dutch shipbuilding firm for two submarines was a significant step towards the modernization of naval capabilities, if only because it is thought to have included the transfer of technology and parts to begin indigenous production of submarines.[48]

Overall, Taiwan has a very solid and extensive naval production potential, based on infrastructure developed since the 1950s for repair and assembly of US naval vessels. Technical competence in this area can, in principle, be transferred to more active design and production work. However, there has been very little innovation in naval vessel development in Taiwan.[49] Substantial resources were devoted in the early 1970s to the development of a civilian shipbuilding capacity, to build oil tankers and other large ships for export. Given the recent decline in world demand for ships, this sector suffers from excess capacity and could be utilized more fully for military purposes. In recent years, the government-run Taiwan Machinery Corporation has made some strides in producing ship components, such as steam turbines and generators, thus strengthening the production infrastructure.

The pattern of production in the naval sector has been influenced by two major factors: (a) the traditional priority accorded ground munitions, aircraft and missiles in Taiwanese defense strategy; and (b) the ready availability of relatively inexpensive surplus US warships in place of new ship purchases or production. In the past, this second factor permitted the Republic of China to give higher priority to non-naval force modernization, but the picture is likely to change, along with Taiwanese defense strategies.

GROUND FORCES

Aside from very limited subcontracting to private firms for small components, all ground-force equipment on the island of Taiwan was, until recently, produced under government auspices by the Combined Service Forces (CSF), the production arm of the Ministry of Defense. The relevant CSF bureaux in the ground-force sector are the following.

Military Industrial Service (MIS)   The MIS is charged with the manufacture and development of all ordnance, including weapons, ammunition, tactical radios and related electronics. The stated objective of the MIS is to make the Republic of China wholly self-sufficient in the production of ordnance and spare parts for US equipment (especially those for which parts or ammunition are no longer securely available) as well as new, more advanced weapons of Taiwanese or partial Taiwanese design. Among the equipment produced in the arsenals under the direction of the MIS are small arms and small arms ammunition, rocket launchers, mortars, hand grenades, artillery projectiles, fuzes, explosives, propellants and naval ordnance.

Military Vehicles Production Service (MVPS)   The MVPS produces at least two types of wheeled military trucks, and has approached complete self-sufficiency in the production of transport vehicles. Although no tanks are coproduced, the MVPS did oversee a program for upgrading M-48 tanks with the US. This production capability has been evolving since 1966, when Taiwan signed a contract for the production of all-purpose transport vehicles.

Quartermaster Service (QS)   The QS plans and produces all military uniforms, gas masks and related materials.[50]

With US help, Taiwan has developed an impressive capacity for small-arms ammunition production. In the mid-1970s, Taiwan purchased elements of the US Small Caliber Ammunition Modernization Program (SCAMP), a high-technology production machinery system for 5.56 mm and 7.62 mm ammunition. A significant surplus of that ammunition is produced in this and other programs.[51]

Under US licensing arrangements, Taiwan has also developed the capability to produce M-14 and M-16 rifles and M-60 machines guns, as well as four calibers of mortars and two types of recoilless rifles. Artillery production includes 125 mm multiple-tube rocket launchers, mounted on US M-113 armored personnel carriers, and 3.5 inch and 66 mm rocket launchers. In addition, Taiwan produces 105 mm and 155 mm howitzers. An explosives and propellants industry, under the aegis of a government arsenal, has evolved to supply the ammunition producers, but the exact quantities of production are not known.[52] Anti-tank capabilities have been improved by a coassembly program for the Rockeye cluster munition, initiated in the late 1970s. Although participation by Taiwanese engineers in this production program is still limited, increased production experience with this munition could, over time, be of considerable importance for the island's defense needs and export potential.[53]

Taiwan has a fairly urgent requirement to improve its inventory of armored vehicles. While planners have attempted to procure the M-60 tank for the United States, they are also endeavoring to develop an indigenous tank production capability. This is a high priority for the army at least.[54] A series of related programs have been undertaken in recent years, including an effort within CIST to develop a laser-range designator, while the MVPS has worked to improve tank frames and stabilizers. The Republic of China has overhauled several thousand tanks since the 1950s, and has superior capabilities in this area. It is reputed that its mechanics can completely disassemble an M-48 tank and restore it to perfect condition. This is a strong base upon which a tank development program could be built.[55]

The condition of the ground-force equipment sector is rather mixed from the standpoint of efficiency. There is considerable overproduction – and excess capacity – for ammunition production and small arms (such as the M-16), as well as in the production of general-purpose vehicles, for which $122 million has been invested since 1966. This has resulted in the production of some 18,000 vehicles, a large number for a country with so few roads.

SOUTH KOREA

National development plans in South Korea have accentuated military industry since the early 1970s. Prior to that time, South Korea had more than one government arsenal producing ammunition and small arms. However, the defense industry as a whole did not command top priority until after the announcement of US troop withdrawals in connection with the 1969 Nixon Doctrine. In 1973, during the annual US–Korean Security Consultative Meeting, the United States pledged formally to assist South Korea in developing its munitions industry. This was to be part of the compensation for reducing the US military presence. Among the earliest Korean–American agreements for the production of weapons under license were a 1971 contract for M-16 rifles, a 1974 contract for M-60 machine guns, and a series of agreements in 1975 for ammunition (20 mm and 50 caliber), howitzers (105 mm and 155 mm) and the Vulcan air defense gun system. Discussions of Korean coproduction of a fighter aircraft were under consideration as early as 1974. All these programs are discussed further below.

The policies of the Carter administration gave Korea its greatest incentives to promote the growth of an independent production capability. The Carter administration's troop-withdrawal plan, later cancelled, served as a major impetus for the régime of President Park

Chung Hee to accelerate investment in heavy industries (chemicals, metals and machinery), so as to provide the infrastructure needed for a larger defense-industrial program. At the same time, the Carter administration provided a palliative for the withdrawal policy in the form of a series of coproduction agreements designed to bolster South Korean weapon production capabilities. These exceeded in quantity and sophistication those offered during the Nixon administration.

WILLFUL HASTE

Nevertheless, the defense industry in Korea today suffers from severe problems. Many of these are the result of the haste with which the Park régime moved. Although fairly important domestic, political and economic rationales can be given for proceeding in this manner, the economic and technological inefficiencies engendered by the overly ambitious investment plans are likely to plague Korean defense industries for some time to come.

The rapidity with which industry responded to the task of creating or converting plant capacity for defense was entirely the result of fiscal and ideological incentives offered during the 1970s. A certain euphoria accompanied defense investment: it was almost completely insulated from commercial pressures and insured from failure by government subsidies. The manner in which funds were disbursed to industry permitted recipients to utilize government resources to import expensive machinery exceeding absorption capacity and output requirements, and to use monies allocated for defense production to prop up flagging civilian enterprises.[56]

Although there was a legal limit of 30 per cent on the share of each firm's total production that could be military in nature, this restriction was not strictly enforced. With investors protected from market forces and responding to fairly stringent government pressures to begin rapid production, there was no prior planning or coordination at the industrial level of what was actually required.

Under any circumstances, the rapidity with which investment took place would have outstripped planning capabilities in the government, even if they had been designed to provide adequate oversight. In the event, if projections were made of the demand for particular goods – in a process for estimating requirements akin to that used by the Taiwanese, for instance – these were ignored.[57]

QUALITY CONTROL, MANAGEMENT AND PLANNING

The two basic deficiencies that plague the Korean defense industry in a more immediate operational sense – and that are at least partly the result of the haste with which defense production expanded – are insufficient quality control procedures and weak management, both official and private. To this should be added a Korean proclivity (at least under Park) for 'prestige projects', such as the development of long-range surface-to-surface missiles. While intended to display evidence of Korean determination and capacity to move into advanced weapon development, and to demonstrate independence from the United States, the net result of these projects has been to divert vast resources into programs that enjoyed only a marginal success, while impeding progress of other, more critical modernization.[58]

Quality control problems stem primarily from the overconfidence of Korean planners about their ability to move rapidly into high-technology production. Early success in the use of US technical assistance agreements and data packages, and in reverse engineering, to produce simple munitions and communications equipment, led to the unfounded assumption that the Korean technical base was ready for more complex industrial processes. A US Military Advisory Group (MAG) study in 1978 states that 'even the failures experienced in the prototyping of field and air defense artillery systems have not fully convinced them of the necessity of paying the price to acquire the how [sic] through technical assistance programs from US government agencies or commercial firms'.[59]

Other problems arise from the condition of the machinery sector in Korea, which, lacking engineers with sufficiently specialized skills, was insufficiently developed to serve as the mainstay of the growing industries. Overall, procedures for ensuring adherence to specifications and the routine conduct of inspections and testing of equipment are still inadequate. They therefore pose rather serious problems for the achievement of military-grade quality in production.[60] In very recent years, this has begun to be rectified, at least in the ammunition production sector, partly through the purchase of new equipment and through greater availability of skilled personnel.[61] Korea has adopted measures to increase its human capital. For example, the Korean Institute of Aeronautical Technology (originally formed in 1978) has increased its recruitment of foreign engineers and boosted its efforts to send Korean students overseas for rigorous training. A new facility at Inchon was initiated, stressing training in high-technology areas such as structural engineering, aerodynamics and avionics. The institute will operate a special branch office to oversee quality control in the F-5 production program.

While technical training may overcome some of the problems of quality control and management, the overcentralized structure of decision-making in both the private and public sectors will militate against efficiency.

Many of the defense production decisions taken during the 1970s were initiated by President Park himself, often without benefit or in contravention of advice given him by technical advisors. Park's initiation of two massive infrastructural projects – the steel facility at Pohang and the highway between Pusan and Seoul – against the advice of professional experts,[62] is thought to have contributed to the president's insulation from advisors during the mid- to late 1970s. The fact that the projects were extraordinary successes heightened Park's tendency to distrust professional advice and to follow his own instincts instead. Although the Ministry of Defense, in conjunction with the Agency for Defense Development (its semi-autonomous research and development arm), the Ministries of Commerce and Industry, and the Economic Planning Board, was formally charged with defense planning and allocations, all decisions were, in fact, made in the Blue House. As a result, it is difficult to depict the 'management structure' of the Korean defense industry. In simplified form, it has consisted of a loose aggregate of eighty-five 'designated' defense production firms working with ADD and the Defense Procurement Agency.[63]

At least in theory, the objective structure of the defense planning apparatus in Korea gives considerable latitude to the Ministry of Defense to establish defense priorities and initiate production projects, the research and development for which is to be carried out by the Agency for Defense Development. ADD and the Ministry of Defense subsequently are to select a firm to carry out the production and testing of prototypes. On the basis of the quality of production at this stage, a firm may become a 'designated defense firm', eligible to receive government subsidies and other financial assistance. It is impossible to trace the extent to which this system has resulted in any quality control – either in eliminating some contractors from eligibility for defense work or in overseeing production to ensure that strict standards are maintained. According to one source, the designation of firms for defense work is a highly political process, depending to a considerable extent on each firm's prior access to government officials.[64] From the beginning, ADD suffered from being insulated from the very industrial programs it was supposed to be directing. The flow of information on desired projects, data and technology for these projects, and other planning criteria, were simply not coherent. ADD also lacked both the authority for, and interest in, oversight once a project was underway.

Structured like a Western think-tank and staffed mostly by Western-educated systems analysts, ADD has little in common with the rising class of entrepreneurs undertaking production for defense. The latter, for the most part, are self-made men with little formal education. Many are financially successful as a direct result of the defense boom of the 1970s. ADD's position in the Ministry of Defense, moreover, is ambiguous. Lines of communication have been weak, since under Park virtually all important decisions were made by a small cadre of advisors in the Blue House, if not by the president himself.

Within the firm, managerial problems are legion. For the most part, they result from a Korean tendency toward rigidly hierarchical management structures, a feature of Korean organizations that appears throughout the economy and government.[65] Korean 'protocol' has it that in practice no criticism is directed upward in the hierarchy. Line personnel rarely report problems emerging in production projects; instead, they pass on responsibility for flaws in design or implementation. In the absence of coordinated procedures to allocate responsibility in a more rational way – through the designation of project managers with routine methods for communicating with workers – problems do not surface until it is too late. This centralization of authority affects decision-making throughout Korean society, and appears likely to remain a fairly immutable condition for some time.

In the absence of government planning and coordination – and in view of the possibility alluded to earlier that duplication among industries was partly a deliberate measure – defense firms soon began to compete viciously among themselves to capture larger shares of the market. Although competition was apparently desired by the government, in the hope that it would generate greater efficiency, the result was chronic excess capacity. Firms attempted to become proficient in the production of all components relating to their particular sector, rather than subcontracting. The result was higher overheads and poor quality. One Korean ammunition manufacturer explained that only through in-house production of the entire line of products – from propellants to explosives to bandoliers for bullets – could 'quality control' be assured.[66] This seems to reflect a reluctance to cooperate with other firms, perhaps in part because of a recognition that the output would in fact be flawed, but also because of a desire for greater short-term profits.

PROGRESS AND ITS LIMITS

A number of reforms were introduced two or three years prior to the change in régime. Others are under consideration. There is some evidence that the present government understands the need to rationalize its incentives policy and relax import restrictions for the ultimate good of Korean industry as a whole. The overinvestment in heavy industry is no longer ignored, and reports indicate that the government, with the cooperation and consent of Korea's oligopolistic 'general trading companies', will move into light industry and high technology, relying as never before on direct foreign investment as a means of gaining access to needed technology.[67]

ADD has been supplemented, though not supplanted, by a government agency called the Korean Institute for Defense Analysis (KIDA), patterned loosely after the US Institute of Defense Analysis, a non-profit organization under the aegis of the Pentagon. KIDA employs most of the more talented systems analysts who were formerly at ADD, and enjoys direct lines of communication with the higher echelons of the Ministry of Defense. Among the measures KIDA is apparently considering for reform of the industrial sector are procedures to estimate the demand for defense output. It is also considering a fundamental alteration of the structure of the defense sector whereby a series of prime contractors would be designated by government decree to be the sole suppliers of particular products, and competing firms would either become subcontractors or be forced out of the defense business. Finally, KIDA has proposed that the 30 per cent limitation on defense production be taken seriously, so as to force the conversion of some plant capacity to more efficient and remunerative activity.[68]

Permitting inefficient companies to founder would be politically very difficult, a potential source of embarrassment for the government as a whole.[69] Moreover, there is an apparently chronic tendency in both government and industry to seek palliatives to structural problems instead of working toward long-term solutions, which tend to be difficult and politically sensitive. While industry looks to the Korean government for continued support, the Korean government, in turn, looks to the United States for direct assistance. Many Korean officials believe that the US could assist their country through the following.[70]

(a) Liberalization of US export restrictions on defense products produced with American inputs or technical data.

(b) Use of Korean industries for the maintenance, overhaul, and improvement of US military equipment deployed in Asia.

(c) Purchase by the United States of Korean output to augment the US mobilization base. With much common equipment in the United States and Korea, and given the ailing state of the US mobilization potential, this type of arrangement is stressed to be of mutual benefit.

(d) Investment by private American defense contractors in Korean industry for various cooperative ventures, including coassembly and coproduction, as well as purchase by the United States of specified products produced by Koreans. Low-cost Korean labor applied within a US-supervised production program is seen to have the potential for highly competitive exports, for mutual benefit.

(e) Purchase by the US Defense Department of Korean ammunition to stockpile in Korea as war-reserve supplies. This stockpile is said to be presently below specified 'required levels' as a result of shifts in Korean government purchases from ammunition to high-technology items.

By far the most contentious issue is that of US restrictions on the sale to third countries of Korean-made products containing US inputs or know-how. Segments of the Korean military/industrial sector – private and official – believe that the United States encouraged the development of Korean defense industries, and so has a continuing responsibility for them, as Korea's 'blood ally'.

A major theme is that US restrictions on Korean exports serve only to curtail Korean trade receipts while arms sales by other countries, including the United States, proceed unabated. Industrialists insist that US commercial interests are the major motivation for the export restrictions, despite the fact that Korea's key competitors tend to be smaller producers such as Italy, Spain, Belgium and Eastern Bloc countries. Korean businessmen also occasionally lament their country's official policy to export only to pro-Western countries, while other suppliers (such as North Korea) are capturing potential Korean markets.

There was controversy in March 1982 over the proposed sale of howitzers, mortars and other munitions to Jordan by the major Korean ammunition manufacturer, Poong-San Metals. The US denied the sale on the grounds that the equipment was partially American in origin and the weapons were in reality intended for Iraq. The logic was not compelling to Poong-San representatives, who stressed that another supplier would simply provide Jordan (or Iraq) in Korea's place. Moreover, it was argued, Poong-San Metals had refused other potential customers – such as Iran – and lost sales in its effort to be 'pro-American'; hence, reciprocity in American policy was justified.

Korea has exported defense products containing US technology in violation of US restrictions. One major example was the export of jointly-produced patrol boats to Indonesia. The Korean government asserted initially that the ships were commercial vessels, but drew back, in the face of overwhelming evidence, to the position that the extensive Korean modifications to the original equipment made them indigenous products, not subject to the terms of the licensing agreement. The same basic script has been replayed in less obvious ways for several years.[71] The stand-off is not likely to be soon resolved.

Quite apart from the merits of Korea's stance on the re-export issue, we may doubt whether the measures proposed for rectification of the country's problems are likely to be adequate to the task. Liberalization of Korean exports of ammunition and other small arms would not make a major impact on the condition of the Korean industrial base. Global demand for these types of defense products is likely to remain constricted over the long term. The number of suppliers alone – including all the industrial countries and perhaps as many as thirty developing nations – will ensure that export potential is limited. Since domestic demand is sufficient to absorb production capabilities in just a few of the exporting nations, Korea faces a permanently glutted global market. Only at the margin could export sales make a difference. To companies suffering up to 90 per cent underutilization of capacity, even one contract is valued, as it buys time.

Korea's real task is to diversify its industry into more technologically complex forms of production needed for the modernization of the Korean armed forces and suitable, at least on a limited scale, for developing export production. A number of factors limit the country's prospects in this area in the near future.

Movement to more competitive or needed defense production has been hampered not only by the diversion of defense resources into the construction of idle plant capacity, but also by investment in high-technology programs that did not succeed. As alluded to earlier, for example, the Park régime was particularly committed to developing a long-range SSM of Korean design. The investment may have resulted in the development of a series of prototype missiles by ADD, only two or three of which apparently entered production. These programs were not based on service requirements, and caused some controversy within the military because of their adverse impact on other modernization goals. Unless it generates technological spin-offs to industry in the form of successful programs, the missile effort will prove to have been no more than a delay in the modernization of Korea's defense industry.[72]

The fascination of some decision-makers in Korea with expensive high-technology defense items creates a serious planning problem. With a defense budget that already represents more than one-third of total government expenditures, and with a very complicated and urgent set of defense modernization requirements, the allocation of scarce resources to politically prestigious but militarily inefficient systems is a serious problem. The history of Korean efforts to acquire the US F-16 advanced fighter aircraft is a case in point, and is discussed in some detail in Chapter 5.

Future modernization programs in Korea – in both production and acquisition – will be hampered by budgetary constraints arising at least in part from past defense spending. A series of planned programs may have to be cancelled for financial reasons. The Korean government has requested a modernization loan from the United States of about $1.5 billion to be advanced over the next five years. The Reagan administration has recommended $210 million in foreign military sales credits for fiscal year 1983, which, though an increase of $44 million over the total for fiscal year 1982, will result in a shortfall in Korea's budget. Most serious of all is the country's state of external indebtedness. The Korean government is paying interest on foreign debts equal to the size of its defense budget and, while it received $166 million from the US in 1982, it simultaneously repaid $254 million in past debts. Domestically, Korea has imposed a defense sales tax of 18 per cent to try to raise additional monies for military modernization.[73]

In the sections below, the individual sectors of Korea's defense industries are described.

AEROSPACE

The most important aircraft coproduction undertaken by Korea is the 1979 agreement for F-5 fighters. Although approved under the Carter administration for F-5E/F models, the program may phase into production of the newer and more capable F-5G – at least, this is the plan that has been proposed to the Korean government by the Northrop

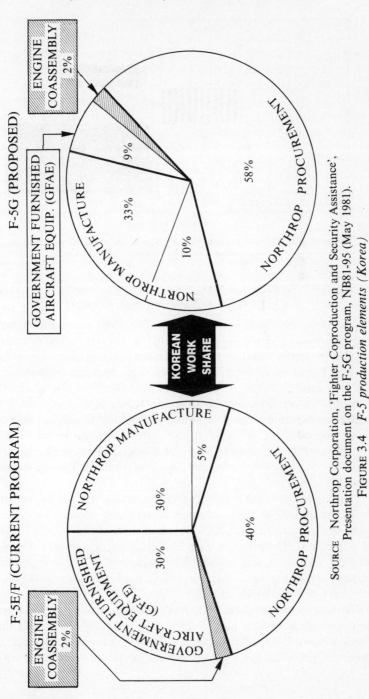

SOURCE Northrop Corporation, 'Fighter Coproduction and Security Assistance', Presentation document on the F-5G program, NB81-95 (May 1981).

FIGURE 3.4 *F-5 production elements (Korea)*

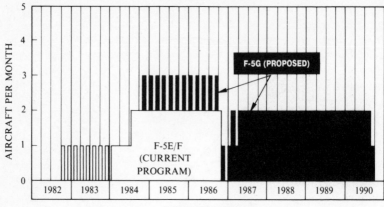

SOURCE   Northrop Corporation, 'Fighter Coproduction and Security
Assistance', Presentation document on the F-5G program, NB81-95
(May 1981).

FIGURE 3.5   *Proposed KAL F-5 production rate/delivery schedule
(68 F-5E/F and 100 F-5G aircraft)*

Corporation (see Figures 3.4 and 3.5). The Koreans originally had
sought coproduction rights for the F–16, but this was turned down by
the Carter administration. Having decided to coproduce the F-5s,
however, Korea has also received approval from the Reagan administration to acquire 36 F-16s. Although coproduction is not part of the F-
16 contract at present, some planners in Korea believe that the F-5
production experience could serve as a sufficient technological base to
enable coproduction of the F-16, were this to be approved by the
United States. Technical and policy problems aside, however, budgetary
considerations may undermine any prospects for this in the foreseeable
future.[74]

The original contract for F-5E/F coproduction was for 68 aircraft
(36 F-5Es and 32 F-5Fs) estimated to cost $62 million. Deliveries
began late in 1982 and will stretch through 1986 (or to 1989 if F-5Gs
are purchased). The program includes full logistical support, training,
technical assistance, and tooling for production.[75] The new aircraft
will replace the existing inventory of F-5As and F-86s, with the possible
retirement of some F-5Bs as well.

The first major program of Korea's aircraft industry, apart from
the construction of four simple trainers in the early 1970s, was the
coproduction of 500 MD helicopters in 1976. The terms of the original
agreement provided for 75 light, combat helicopters (equipped with

2.75 inch rockets and machine guns) to be assembled from kits provided by Hughes Aircraft. Assembly of the 75 helicopters was completed in March 1979, and it was then reported that another 48 would be built at a facility in Kimhae.[76] The program did produce certain technological spin-offs to the civilian sector. At least six civilian versions of the helicopter were delivered to local companies like the Korea Tacoma Company, with plans for a similar number to be delivered in later years.[77]

With some early difficulties, the coproduction program as a whole moved through successive stages of technical complexity, from assembly of kits to production of major components.[78] On the basis of its success, a joint US Navy and Air Force contract was awarded to the Hanjin Corporation, a subsidiary of Korean Air Lines, for the maintenance and repair of F-4D/E aircraft deployed through Asia. (This service had been performed by Air Asia in Taiwan, but the arrangement was terminated following the normalization of relations with mainland China.) About 30 aircraft were maintained by the Koreans in 1980. The process is quite extensive, and includes the disassembly, inspection and overhaul of aircraft, as well as some upgrading of components. The entire process takes about four monts, and is required for all F-4s every four to five years. New facilities were built in Korea for the purpose (4.4 million square feet in the first year, and expected to double). The contract also resulted in more extensive training programs for Korean technicians by the US Air Force.[79]

Although Korean Air Lines is the prime contractor for all aircraft production, some work on engines is done at Samsung Precision Industries. Samsung is working on turbine engine technology, hoping to assemble engines and produce some turbine components. Increasing the number of firms involved in aircraft projects has been a declared objective of the Koreans since the mid-1970s. An Aircraft Industry Bill passed in 1978 to formalize this objective, for instance, called for production of aircraft by 1990 from components and equipment made in Korea.[80]

Equally important, at least under the Park régime, was the development of an indigenous capability for missile design and production. The earlier related program was a MAAG-supervised maintenance facility for the Hawk and Nike Hercules missile systems, begun in 1972. It was recognized at the time by the US government that the location of such a facility in Korea 'could save several million dollars over the life cycle of these missiles given lower labor costs, reduced transportation costs, and utilization of local repair parts'.[81] This program led to the establishment of a commercial maintenance facility in Korea, under the aegis of a local firm, Gold Star Precision Industries, Ltd. Korean personnel have received training from the US military and, among others, from the Raytheon Corporation, to continue making improvements in this commercial facility.

Domestic missile R&D and production were undertaken, at least until quite recently, within the Agency for Defense Development. President Park staked his personal prestige on the development of missile production capabilities, with special emphasis on the development of a long-range SSM capable of reaching Pyong-yang.[82] The latter program, conducted in a clandestine manner against the advice of the United States, caused protracted controversy between the two countries for several years. Given the limitations of Korean guidance technologies, this SSM could have been militarily effective against such North Korean targets as airfields only if it were connected to the development of a chemical or nuclear warhead. There is some evidence to suggest that the Park régime did have a fairly extensive program aimed at developing the latter. However, it should be noted that interest in developing a conventional SSM seems not to have hinged on demonstrations of strict military utility.[83]

South Korea tested its first SSM successfully in 1978. The missile is believed to have a range of 100–160 km. The SSM is designated NH-K and is a modification of the US Nike Hercules, although it is presented domestically to have been made entirely with indigenously produced components.[84] A military display in mid-1979 revealed three Korean-developed missile systems, although it is not possible to determine the extent to which these relied on US technical data. They include a multimissile launch system with 28 launching tubes, a lightweight anti-tank weapon, the NH-K described above, and another, smaller SSM said to resemble a shortened version of the US Honest John, launched from a trailer.

ADD gained experience in missile technology in part through a program for maintaining the US Nike Hercules system. Under the agreement, the United States provided technical specifications for upgrading reliability through electronics improvements, improving conventional warheads, and adding a capability for use in a surface-to-surface mode.[85] The latter may have resulted in considerable augmentation of the missile's range, although it has not been possible to verify this. In principle, the new guidance systems should increase range. In any event, one analyst has suggested that this system could be turned fairly easily into 'a fine tactical nuclear weapon'.[86] In addition to the NH-K, Korea has had a program to modernize the Honest John missile by adding improved guidance systems to increase its accuracy.[87]

NAVAL PRODUCTION

Korea built its first successful naval vessels – fast patrol craft – under US supervision in 1972.[88] Joint US–Korean production of four multimission patrol ships armed with Harpoon missiles followed in 1974. An additional eight armed with Exocet missiles were produced in 1979–80; four more were ordered in 1982. The purpose of naval production in Korea is the prevention of North Korean insurgent operations on the South Korean coast. As a result, the country has concentrated on the production of high-speed patrol boats for guarding shallow inshore waters. Various landing craft, including four capable of carrying tanks, have also been produced.

Eleven corvettes are in the process of construction. These are equipped with a Dutch weapon-control system, US–Italian gas turbines and West German diesel engines. Armed with one 76 mm and two 30 mm guns, the corvettes may be fitted with electronic warfare equipment if testing proves successful.[89] A prototype submarine was developed in the late 1970s, but the technical ability to manufacture a modern submarine in Korea will be lacking for some time to come. The South Koreans have also built a guided-missile frigate, but do not seem to have the capability to build larger naval craft. A significant weakness in the naval area has been the difficulty of maintaining existing inventories. Since US destroyers and frigates of Second World War vintage are the mainstay of the Korean Navy, spare parts are increasingly difficult to obtain. Naval readiness will suffer if these ships are not replaced over time with more modern equipment.[90]

Korea's major naval producer is Hyundai Shipbuilding and Heavy Industries, which has a massive shipyard near Ulsan, in the southern part of the country. It is reported that Hyundai is in search of overseas export markets and has had discussions with various Latin American and African governments in recent years. Another firm, Korea Tacoma Marine Industries, built the patrol boats exported with such controversy to Indonesia. Although the country has the capability to produce basic naval craft, major weaknesses include engines, naval armaments and components such as fire control systems, which must be imported.

GROUND FORCES

Most Korean production programs are concentrated on ground-force equipment, ranging from armored personnel carriers (APC), tanks and fairly sophisticated ordnance design and production, to simple munitions. About half of Korean needs are met by domestic production; the great majority of this is intended for use by ground forces.

Upgrading the M-48 tank is a major program, in which the M-48 is standardized to the M-48A5 configuration of US army tanks deployed in Korea. This package includes a 105 mm gun, a standardized engine, fire-control equipment and night-vision devices.[91] Korea has wanted to produce a tank domestically since the mid-1970s. In 1976, extensive discussions were held between the South Korean Joint Chiefs of Staff and the US Military Assistance Advisory Group mission to consider options for improving Korean armored forces. Korea pressed for acquisition of the M-60A1 tank, but was denied. For the M-48 project, granted in its place, South Korea reportedly produced up to 30 per cent of the conversion components.[92]

Korea's ambition in the tank field has come a step closer to fulfillment under a recent agreement with General Dynamics for the production of 'several hundred' main battle tanks. Designated ROKIT, for Republic of Korea Indigenous Tank, the new tank carries a 105 mm gun and is reported to be an upgraded M-60 similar to that which Korea sought unsuccessfully in the late 1970s, or an 'austere version' of the M1 Abrams. Production in South Korea is due to begin in early 1985, after prototype testing which began in April 1984.[93]

Another upgrade program is being undertaken by Daewoo Heavy Industries at Inchon to improve the Korean army's Second Division fleet of M-113 APCs. In the future, Daewoo is to have the responsibility of performing routine depot and maintenance work for all army tracked vehicles, so that they do not have to be sent to the United States for overhaul. Korea has also received assistance from Italy in this area. An agreement for coassembly of the Fiat 6614A four-wheeled APC was signed in 1976.[94]

South Korea also produces military trucks. KIA Industries (a merger company that includes the former Asia Motor Corporation) was designated the major contractor for all wheeled vehicles produced domestically for the military. One program calls for the insertion of a Korean-made four-cylinder engine in military jeeps, while larger trucks are being retrofitted with a Korean-produced German diesel engine. A quarter-ton truck of KIA design was under development in 1978,

Item	Comment
M-16 rifle (5.56 mm)	Produced under license with Colt Industries and the US government
M-14 rifle (7.62 mm)	Estimated 100,000 produced by 1982
M-60 machine gun (7.62 mm)	US design
155 mm howitzer	US design
105 mm howitzer	US design
8 inch self-propelled howitzer	US design
81 mm mortar	Permitted for export by US
60 mm mortar	No longer produced
4.2 inch mortar	US design
106 mm recoilless rifle	Permitted for export by US
90 mm recoilless rifle	US design
Vulcan anti-aircraft gun	US design
Oerlikon anti-aircraft system	Swiss coproduction; exports have to be sold through the Swiss who take 50% of sale revenues
M-19 anti-tank mine	US design
M-18 anti-personnel mine	US design
M-72 rocket launcher	US design
M-203 rifle grenade	US design
90 mm armor-piercing projectile	US design

SOURCE Defense Market Survey, Inc., *DMS Market Intelligence Report for China (Taiwan) and South Korea* (Greenwich, CT: Defense Market Survey, Inc., 1981).

FIGURE 3.6 *Ammunition and ordnance production in South Korea*

and plans called for the design and production of a two-and-a-half ton truck by the end of 1982. Korea also has a contract with American Motors General Products for the assembly of a five-ton truck (M-809 series) using the Korean-produced diesel engine.[95]

Korea produces a full range of ordnance, artillery and ammunition, as well as propellants and explosives (see Figure 3.6). Prototypes for the 105 mm and 155 mm howitzers were manufactured in Korea as early as 1973 and 1974, respectively, from US technical data packages (TDPS). ADD simply copied the TDPs and redrew the designs to Korean specifications without US assistance, providing them in turn to Korean firms.[96] As defects increasingly surfaced in manufacture of these weapons in the mid-1970s, the Koreans requested further technical assistance, which was provided by personnel from American arsenals. Prior to that time, severe malfunctions affected indigenously-produced howitzers and Vulcan anti-aircraft guns. These problems stemmed largely from the haste with which the Koreans initiated production programs. Since then, manufacturing capabilities have improved commensurate with efforts to abide more closely by US specifications and to perform appropriate tests during the manufacturing process. The quality of small-caliber weapons and ammunition in Korea is reputed to be very good, while weapons of larger caliber and components are approaching military-grade quality. The improvements have much to do with improvements in training, although shortages of both skilled inspectors and modern testing equipment are still in evidence today.[97] Among the more advanced development projects in this sector are the testing of new, small tactical rockets and the retrofitting of laser range-finders for the Korean tank force.

The munitions sector in Korea is suffering most acutely from excess capacity. The Poong-San Metal Manufacturing Company (PMC), the major ammunition and ordnance producer in Korea, is in some areas 90 per cent idle. The company recently purchased a government arsenal (which produced 5.56 mm and 7.62 mm ammunition rounds and M-16 rifles, among other items) offered for sale to commercial firms because of its heavy financial losses. PMC bought the arsenal on fairly ruinous terms (including a stipulation that no workers could be displaced for two years) in order to prevent competitors from doing so, a prospect that was thought to have the potential to undercut what was left of PMC's market. The purchase of the arsenal, coupled with excess capacity in other PMC defense plants, has placed the company in a perilous position.[98] PMC has exported in the past to the Philippines, Thailand, Malaysia, Guatemala and Cameroon, but lately has found few export customers, and was strongly affected when the United States failed to approve the sale of howitzers and mortars to Jordan.[99]

The electronics industry in Korea has achieved some advances in recent years, especially in the area of data processing, communications and computers. Electronics are a major focus of the most recent five-year development plan. The government has developed a state-run organization, the Fine Instruments Center, to oversee advances in electronics, machinery and related industries. This is a major area and is targeted for expansion, in part through the initiation of research institutes with special emphasis on industrial electronics such as semiconductors and communications equipment.[100]

INDIGENOUS MILITARY PRODUCTION AND ALLIANCE POLITICS

There is a double advantage to building up a domestic capacity to produce sophisticated weaponry. Having such a capacity *in potentia* can make it easier to command technical and material assistance that the leading nation in an alliance, ever concerned about keeping its lesser partners in their prescribed place, might otherwise refuse to give. However, that very same production capacity, whether or not fully exploited, may heighten tensions with states outside the alliance, particularly if it is viewed by regional antagonists as a portent of future belligerence. The development of indigenous defense capabilities – for offensive, defensive or export purposes – is an understandable expression of the strong urge to self-reliance in the maturing state coming of age in the modern world. The process nevertheless produces inevitable stresses within the alliance and the region to which the developing country belongs.

Korea and Taiwan have utilized their ability to develop certain types of weapons to gain concessions from the United States. The pattern of arms shipments to Korea following the friction over that country's long-range missile development program demonstrates that the United States did augment its transfers as a means of dissuading Korea from pursuing its missile program. Shortly after the diplomatic exchanges concerning Korea's clandestine missile development efforts, the United States decided to provide improved capabilities for the Nike Hercules and Honest John missiles already in the Korean inventory.[101] Some also suggest that the United States became more forthcoming in providing Korea with technical assistance for other types of missile development, increasing consultations between US and Korean military planners on Korea's missile requirements, for example.[102] On balance, the threat by Korea to develop a long-range missile thought by the US to be potentially destabilizing proved of benefit to Korean military modernization. Left to its own resources, Korea might have failed in its indigenous effort.

Although there is less concrete evidence of successful use by Taiwan of leverage obtained from weapon development programs, it has been the clear policy of the island's leaders to put into development – but not production – types of technologies needed for force modernization efforts, but to do it so as to induce the United States to sell them the equipment instead. Purchase avoids the high risk and cost of putting into production systems which are not yet fully developed or refined and for which domestic demand may be too limited to make production economical.[103]

Taiwan and Korea may one day reach the point where the decision to develop particular technologies considered by the US to be destabilizing becomes an entirely internal matter. It is thought by some sources that Korea and Taiwan are already close to being able to deploy nuclear weapons.[104]

If an ally has a latent ability to produce a type of advanced weapon thought to be regionally destabilizing, the United States may attempt to halt such development by promising concessions or threatening to invoke punitive sanctions. In the nuclear area, concessions have generally taken the form of additional transfers of conventional arms, as was also the case in the exchanges over indigenous developments of conventional missiles. A variety of punitive sanctions are also available, ranging from termination of certain types of military assistance to the interruption of overall alliance support. These latter measures will vary in effectiveness from country to country.

Given the extreme fragility of Taiwan's security position since derecognition, the continued, albeit limited, support of the United States is absolutely critical. Taiwan must calculate very carefully how to use any military leverage it may derive from its technological capabilities. On the one hand, Taiwan's nuclear potential seems to have helped the country to secure continued US support. This is understandable, in view of the US commitment to non-proliferation and the gravity of the probable effects if Taiwan were to explode or reveal that it possessed a nuclear weapon. This is also true to a lesser extent for conventional weapons. The United States wishes to see that Taiwan does not develop a long-range surface-to-surface missile capable of hitting the mainland. To accomplish its goal, the United States may have provided Taiwan with needed technologies for other types of weapons. As long as bargaining room remains, such tactics may benefit Taiwan. Once Taiwan actually deploys a disputed weapon, the picture changes drastically. In itself, this limits the bargaining power of the island's leaders, for the risk that the United States might in fact impose punitive sanctions is simply not one that Taiwan can afford to take.

Korea enjoys a more commanding position by virtue of its far greater interdependency with the United States in security and defense issues. The importance of the Korean peninsula to continued stability throughout North-east Asia is such that the United States could not afford to permit a serious loss of influence over Korea or to allow the country to undertake destabilizing actions of various kinds. Punitive measures against that country must be weighed carefully against the risks that Korea will, in fact, continue to act in a fashion inimical to US interests.

Needless to say, Korea, like Taiwan, has reasons to avoid unduly testing its relationship with the United States. Its defense capabilities, including production capabilities, are critically dependent on US inputs for continued viability, and the country could not afford even a temporary interruption of this source of supply. Both countries, moreover, depend upon the United States to reinforce their national legitimacy in an international environment where this legitimacy is challenged constantly.

Generally, the defense industries of Korea and Taiwan have not yet had a significant impact upon the existing forces in the two countries. It is, therefore, not surprising that Korea and Taiwan have refrained from altering their fundamental defense strategies as a result of the weapons they can produce. At the present stage of both industries, the importance of production capabilities is still more political than military. Defense production represents a *process* which will become militarily significant in later years. What does this imply for the two countries' regional relations and relations with adversaries?

Were technological capabilities to be used to achieve 'breakthroughs' that permitted qualitative leaps in the type of weapons deployed in the region – long-range SSMs in Korea, for instance – this could complicate the regional security system upon which a fragile stability has been based. For Taiwan, the situation is somewhat dissimilar, given the tremendous disparity in the size of its forces relative to those of the PRC. Nevertheless, regional political tensions do arise as a result of direct transfers of highly visible items from the United States. The PRC is trying to terminate such transfers as part of its efforts to isolate Taiwan from continued US support.

For the time being, improvements in the defense sectors of Taiwan and Korea are likely to continue to consist of incremental achievements in production capabilities for what are still fairly rudimentary weapon systems. Were conflict actually to break out in the region, weapon production capabilities for even simple munitions could make an impact on the balance of forces. They would ensure a source of ready supply of certain weapons without requiring the sanction of the United States. However, the long-term supply of weapons for even a simple conventional confrontation would not be assured, as the US still controls certain critical spare parts needed to keep production lines in existence.

As Taiwan and Korea move into production of more sophisticated weaponry, arms exports will become ever more important. Varying perceptions of dependence on the United States account for differences in the degree of independence shown in this regard. Korea has tended to be fairly loose in its interpretation of American export restrictions, and has had fairly serious disagreements with the United States over sales of equipment produced with American licenses or technology. The instances of Taiwan's violations of American export guidelines, by contrast, have been negligible. Again, Taiwan has feared losing American support and hence has seen a need to be more circumspect. The potential for exports from these countries, however, is not significant. In fact, the difficulties they have in marketing military technology that is competitive with the products of industrial countries suggest that Taiwan and South Korea may have to seek market outlets wherever they can – in other words, countries which for political reasons the United States or other Western states do not support militarily.

It seems that as both countries move towards greater technical proficiency in weapons development, they will have at least the option of exercising greater autonomy in certain areas of defense policy. Continued dependence on the United States aside, the capability to develop, produce and deploy weapons which violate US notions of desirable defense materiel will bring in its train diplomatic bargaining power which can be of benefit to the producing country. At the base of this dynamic is a contradiction in US policy on the subject of arms and technology transfers to these states. On the one hand, the United States seems to favor self-reliance among recipients, and carries this out by providing weapons technology and technical assistance to aid states in developing production capabilities. On the other hand, it seems never to have doubted that it would continue to have control over the disposition of defense resources, with the concomitant authority to ensure that these were used in a moderate, defensively oriented fashion. Once technology is transferred, however, the US position is not so favored. The transfer of production technology is not a reversible act. It cannot be recalled at will, regardless of the transgressions the recipient may commit.

SUMMARY

This overview reveals important differences in the manner in which Taiwan and South Korea have approached military industrialization. As will be seen in later chapters, these differences presage far broader disparities between overall development strategies in the two cases.

For reasons made clear in the next chapter, Taiwan's defense-industrialization strategy reflects the tremendous importance its planners place upon maintaining economic stability and efficiency. The country has accordingly chosen carefully among defense production projects and to the extent possible has encouraged linkages among defense and civilian projects so as to maximize positive spin-offs to the civilian economy and avoid economic dislocation. Taiwan's planners have also focused extensively on the development of a scientific infrastructure to serve as a focus for human-capital development and technological innovation, and have encouraged only selected, efficient firms to undertake defense production projects.

Korea, by contrast, made great changes in its industrial structure on behalf of defense production, but did so with little regard to traditional criteria of economic or military efficiency. Both industrial and military modernization have suffered from the mismanagement and depletion of resources associated with the Park régime's overly ambitious defense projects. Still lacking an even partially adequate scientific infrastructure, the country is saddled with a collection of over-capitalized industries that from the outset have placed profits ahead of standards of industrial quality. It is important, however, not to paint an overly stark and Manichaean portrait of the differences between the two cases. The factors underlying each state's threat perception, economic strategy and political ethos, the subject of the next chapter, should better explain the sources of these disparities.

4   Modernization in the Garrison State

THE ECONOMIC OUTLOOK FOR TAIWAN AND KOREA

Taiwan entered the 1980s with guarded economic and political optimism. The new administration in the United States promised a much improved international climate for continued economic and military modernization. In spite of disappointments in the geopolitical sphere, Taiwan's economy has continued to prosper with steady, if slower, growth, and has sustained expectations that, barring any new, drastic realignments in the international political system, high levels of foreign investment and trade will permit continued gains in coming years.

Korea's prospects in the 1980s are more difficult to decipher than Taiwan's. Most of Korea's difficulties are internal, more the result of government policy than of limits on the country's resources or capabilities.[1] The interplay of economic and political forces is so entwined, moreover, that it is difficult to assess the role that purely economic considerations will play in Korea's future economic evolution.

Taiwan

Three major features of the current Taiwanese economic climate merit emphasis: (a) the continuing need to shift the economy away from labor-intensive sectors to technology- and skill-intensive industry such as computers and advanced machinery, in order to keep pace with the changing pattern of market forces in the international economy; (b) the continued development of capital-intensive heavy industry at an incremental and selective pace in order to sustain economic modernization without interrupting high-growth economic activity; and (c) the expansion of trade relations with traditional and new trading partners in order to sustain the level of exports and secure imports necessary for development, especially imports of energy and advanced technology, including military technology.[2]

Although the new development plan stresses these three areas of priority, constraints on continued high growth and persistent sectoral imbalances, both internal and external, may make modification necessary. Recession is causing slower growth in exports and investment, and lower corporate profits. Individual sectors are experiencing specific growth problems, including petrochemicals (largely because of international competition), agriculture (partly due to a drought in early 1981, followed by floods) and construction (due to high interest rates). Lower demand at home and abroad has affected the manufacturing sector. Aluminum production fell by 31.3 per cent in 1981 and required government intervention to keep the state-owned Taiwan Aluminum Company viable. Iron and steel have also slumped due to a global downturn in construction.[3]

Despite slower exports and high energy costs, Taiwan maintained a rate of 6–7 per cent annual growth between 1980 and early 1982, lower than the official targets of 8 per cent but still quite solid. In spite of the recession, the Taiwanese government continues to emphasize the shift toward greater capital- and technology-intensity in industry. Technology-intensive projects center on the Hsinchu Science Park, where firms have been established to develop minicomputers, integrated circuits, microprocessors, precision machinery and turbine generators. The defense component of these projects is both derivative and direct. The activities of Hsinchu firms are of clear relevance for defense-technology production in the 1980s.

In summary, over the next ten years Taiwan will continue to develop its industrial infrastructure so as to be able to produce quality products competitive on the international market. Economic stability will require domestic policies that stem the inflationary pressures arising from energy imports, higher wage scales, and the costs of building a base for heavy industry. In both civilian and defense production, a major objective will be to secure advanced technology and technical assistance from abroad to aid in these endeavors. In areas related to defense, this may be particularly difficult. Nevertheless, Taiwan will explicitly seek to increase compatibility between economic (profit-making) and defense activities, in order to maximize the efficiency and development of both.

Korea

Korea was plagued by political unrest and turmoil in 1979–80, culminating in the assassination of President Park and the accession to power, after some dely, of General Chun Doo Hwan. The political turmoil was closely associated with economic difficulties during this period, and the government's reaction to them. The two major problems were high inflation and a huge balance of payments deficit due to a decline in the export market, increases in import prices, and years of deficit financing.[4] Korea's microeconomic difficulties were even more daunting. They consisted of chronic excess capacity in industry, continued depletion of government financial resources from desperate attempts to keep companies afloat, an increase in the unemployment rate to 6 per cent, and declining investor confidence, domestic and foreign, that threatened to provoke a flight of capital from Korea.[5]

The government adopted measures to rectify these problems, but they produced unintended adverse effects. To stem unemployment, the government made it illegal to fire workers. This added severe burdens to industries that were already struggling with inflated prices for production inputs. The government effected a 20 per cent devaluation in the won, and then permitted it to float. This raised the dollar exchange ratio by 34 per cent, compounding the difficulties of firms dependent on foreign capital.[6] To curb inflation, the government clamped down on credit to industry, raising interest rates from 3 to 6 per cent. Added to the decline in sales and the burden of retained workers, credit restrictions created severe cash flow problems for some businesses. The government changed this policy in late 1980, in part to prevent bankruptcies.[7]

The major threat to Korean economic and political stability was the decline in investor confidence. The multiplier effect of such a decline can deepen a recession and compound all of the political and economic problems of the original economic conditions. The restoration of political order under Chun, combined with some success in stabilizing prices and restoring export competitiveness, resulted by 1981 in an economy that was again functioning in a predictable fashion. This helped to staunch the flight of capital and contributed to renewed economic growth.

The present challenge to Korean economic growth is more political than economic. The industries which developed under the permissive conditions of the Park régime must retrench considerably from the flexible business practices and dependence on government largesse to which they have become accustomed. Some of these firms, among them many engaged in defense production, will have to close or merge with others. Because Korea's delicate political stability has, in the past, depended on accommodation by government of vested interests, Chun, whose administration has already been branded 'anti-business', will have to proceed very carefully in the rationalization process.

Taiwan and South Korea share a number of economic characteristics that have implications for their long-term security. Both are highly dependent on trade, making their economies extremely sensitive to world economic trends. Both devote enormous resources to military preparedness, at obvious cost to their economies. Both have fairly new technical infrastructures that require constant monitoring and nurturing to sustain growth. The differences between the two states, however, overshadow their similarities. Taiwan, as a result of careful and deliberate policy, has managed the difficult process of economic growth with a minimum of disruption and political discontent. Taiwan's leaders understand that economic difficulties can only heighten the island's political isolation, perhaps at last to a fatal degree.

Korea, by contrast, has been overly ambitious and aggressive, and the economic problems that threaten its long-term security are largely, though certainly not solely, of its own making. With repeated assurances of continuing outside assistance, a luxury not afforded Taiwan, Korea has expanded its economy beyond the ability of domestic structures to accommodate change. Rectifying the serious imbalances that now exist after more than two decades of blind and rapid growth will require major reforms in the Korean economy.

ECONOMIC GROWTH: THE BEST DEFENSE

Since their creation as modern states, Taiwan and South Korea have each managed their economy as 'an integral and vital part of the garrison state'.[8] Decisions in matters of economic development must therefore be analysed simultaneously for their contribution to the military security of the state. At the same time, military decisions – whether or not to produce a given weapon system indigenously, to purchase it ready-made, or to forgo it altogether – must be looked at from the point of view of their contribution to the economic health of the state. In contrast to the traditional analytic model that posits defense spending as a public good almost foreign to the economy, a drag on the more respectable goal of achieving analytically distinct development objectives, we must, when dealing with states whose leaders believe themselves to be under constant foreign or domestic threat, acknowledge that investment for defense and development is highly interdependent. This is particularly true when defense investment is devoted to indigenous military production of a sort capable of creating linkages with the civilian economy. In such cases, the decision to invest resources entails economic planning, guided and informed by the criteria appropriate to such endeavor. It is no longer a simple matter of maximizing military might.

Over time, the interactions among defense and economic considerations in each country produce a particular 'development style'. We shall see that whereas Taiwan has come to be supremely cautious in all matters affecting her economy – in recognition of threat to external and internal stability posed by bumps and jolts in the economic development process – Korea, with the luxury of continued US support, has been far less circumspect in harmonizing military and economic goals. As a result, her economic policies in the postwar years pose greater threats to her ultimate security.

Economic success has permitted both states to maximize security objectives through the ability to secure deeper trade relations with other countries, to ensure access to imports, technology and the expertise needed for modernization (economic and military), and to help stabilize internal and external conditions by providing a foreign presence interested in maintaining the status quo. At the same time, economic success of the type exhibited by both countries – achieved through increased exposure to and interaction with the forces of the international market – has increased the vulnerability of these states to exogenously-induced instabilities. Changes in the international economy, such as a sudden alteration in commodity prices (energy, in particular), in foreign economic or political practices (such as trade protectionism), in politically-induced restrictions on exports of strategic goods (particularly advanced technologies), or simply changes in the terms of economic competition in the marketplace, all have the potential to undermine domestic economic equilibrium.

It is within the broader sphere of economic/defense variables that the national planners of Taiwan and South Korea have had to make decisions about the most efficient means of enhancing their security, designing strategies to reflect a sound mix of economic and military modernization objectives. Investment in defense in preference to other sectors inevitably involves trade-offs among competing investment demands. The assessment of these trade-offs has become an increasingly urgent issue in the current decade. Until the early 1970s for Taiwan and the middle of the decade for Korea, it was possible to sustain high defense budgets without major sacrifices to civilian economic modernization, at least in part because of the continued provision of American grants and aid. Since that time, however, the termination or contraction of American assistance has coincided with higher costs for weapons technology and for production, maintenance and support of weapons, rendering decisions about allocations of scarce resources far more critical.

As was discussed in previous chapters, the difficulties faced by Taiwan and South Korea in ensuring a consistent source of military supply from the United States (in imports of weapons and in US troop deployments) led in the early 1970s to a concerted effort by both states to attain a degree of self-sufficiency in force planning, with the development and production of weapons a major aspect of these efforts. At the same time, this type of investment necessitated economic policies in which rapid growth became as vital to security as was the successful mastery of modern military capabilities. In fact, the two sets of measures are not separable impulses. The essential task in both countries has been to harmonize, to the extent possible, the demands of rising military requirements with the overall achievement of a type of economic prowess capable of adapting to, and recovering from, the vagaries of international economic and political forces.

EARLY INDUSTRIAL GROWTH: 1945–60

Taiwan and South Korea share a legacy of Japanese colonialism. Although we shall not examine this period in detail, it is important to note a number of facts that are particularly relevant to our later discussion.[9] First, a commonality. Japan's colonial policy of restricting education and economic participation was applied to both Taiwan and South Korea. In Taiwan, this deficiency was partially overcome by the influx of a million and a half mainland Chinese by 1949, among them six to seven hundred thousand military personnel and a sizable portion of China's technical and administrative élite.[10] Migrations did not yield the same concentration of skill in South Korea as in Taiwan, if only because Korea did not have a pool the size of China's from which to draw.

A second important distinction between the two states derives directly from the Japanese occupation. Taiwan's colonial development was primarily agricultural, whereas Korea's was somewhat more industrial in character. After 1949 in Taiwan, the incoming mainlanders recognized the ills of the island's typical Chinese agriculture and were detached enough from the Taiwanese social structure to introduce positive changes.[11] In its subsequent economic development, the solidity of the Taiwanese agricultural sector served as a point of political and economic stability for the island, as we shall see.

The two countries are similar in their economic factor endowments. Both countries are extremely poor in indigenous resources, such as minerals. Mining in both countries was negligible even in the late 1960s, accounting for less than 2 per cent of GNP in Taiwan and a totally marginal amount in South Korea. Both countries remain highly dependent on imported energy, a dependence that has grown commensurate with energy-consuming industry. Neither, from an economic standpoint, has any particular geographical or climatic advantages, other than their common configuration as island or peninsula, a fact believed by some to permit 'naturally open economies' beneficial to growth through trade. Both have very high population densities in relation to the amount of cultivatable land, densities that exceed those of Bangladesh, Egypt, and even India.[12]

The willingness of the work force to work long and hard is an economic resource with which both countries are amply endowed. The work ethic exhibited in both Taiwan and South Korea is truly remarkable by international standards, and is a significant advantage for both governments in their efforts to transform the economy. No doubt rooted in Korean and Chinese historical and philosophical traditions, this capacity for concentrated self-application appears 'to be relatively invariant to the nature of the economic system and even to personal incentive'.[13]

However, it is in terms of the differences between the two states that their characteristic 'development styles' can best be analysed. In the remainder of this chapter we shall trace the emerging differentiation of Taiwan and South Korea along four dimensions:

(a) geopolitical place (the relations of each state with the rest of the world);

(b) political culture (including the relative strength of military and civilian institutions);

(c) the degree of government intervention in the economy, and the degree of support given to research and development and human resource development;

(d) agricultural development (the linchpin of balanced growth).

The Importance of Geopolitical Place

Perhaps the single most influential factor in the immediate postwar development of Taiwan and South Korea is the geopolitical place both states occupied at the vortex of international conflicts as profound in impact as they were beyond the respective state's control or influence. Both were catapulted into the mainstream of modern political history by conflict among greater powers. As we have seen, the United States especially came to regard the two countries as strategic assets, and developed a stake in promoting their internal development.

While Korea clearly emerged as a modern state as a product of multinational bargaining and military conflict, Taiwan also has had its geopolitical identity defined by such forces. The enshrinement of Taiwan as the legitimate government of China stemmed solely from the profound enmity between the People's Republic of China and the United States following the communist victory in China's civil war and the Chinese intervention in the Korean conflict. The Korean war is the crucible from which both states emerged as powers considered strategically vital to the United States.

The development of each country has continued to be affected by military conflict involving larger states, ranging from the direct confrontation between Taiwan and the People's Republic over the offshore islands of Quemoy and Matsu, in which US intervention played a critical role, to the profound impact of US military policies in South-east Asia on both countries' military forces. In more recent years, both countries have had to confront challenges stemming from the United States' redefinition of their strategic importance.

The combined effect of military and political conflict in and around Taiwan and South Korea during this period, in summary terms, was to infuse both countries with foreign aid and expertise (in Taiwan's case, altering the population completely through the influx of mainlanders) and to establish economies geared in large measure toward the support of large military establishments. In turn, military forces became a central focus of the development of both states, although the degree to which this was true differentiates the two countries markedly. Whereas in Korea the condition of domestic civilian institutions was so weak as to prompt intervention by the only viable alternative, the armed forces, Taiwan maintained an autonomous civilian government and created a more successful network of cooperative domestic development institutions that worked in tandem with the military.

The influence of foreign aid advisors may have been crucial in directing both economies toward private enterprise and export trade after years of careful protectionist policies aimed at maximizing economic stability. Some would even argue that it was directly the result of American aid and advisors that both countries undertook economic reforms in the early 1960s, moving into international trade, and that in their absence the tendency would have been to 'retreat inside the walls of a seige economy'.[14]

The period of the 1950s and early 1960s for Taiwan and South Korea was characterized by slow, steady growth, sustained in large measure by generous amounts of economic and military assistance from the United States. In Korea, the emphasis until the early 1960s was reconstruction, stressing economic stability over growth. No serious efforts were made to increase domestic savings and investment, and the governments' strategies seemed entirely geared toward maximizing American and other foreign assistance.[15]

Syngham Rhee's régime was pledged to seek reunification with the North prior to any structural change in the South that might be seen to consolidate the status quo. Thus American aid, which exceeded $1.5 billion between 1953 and 1957, focused on developing a minimum of industrial infrastructure, restricted to electric power generation, transportation and rudimentary communication. American military aid, which represented over 10 per cent of the total GNP of Korea in the late 1950s, also provided for a range of support services and construction activities, as well as for direct modernization of the Korean armed forces through equipment transfers. Korea's own expenditures on the military were comparatively low, averaging about 5 per cent of GNP until the mid-1960s, less than half of the burden shouldered by Taiwan.[16.]

US economic assistance to Taiwan was lower than to Korea. It was estimated to have reached about $1.3 billion between 1951 and 1964, the year of its termination. Extensive military assistance was also provided, amounting to an estimated $2.5 billion in the same time period. The latter permitted Taiwan to support a military force that exceeded six hundred thousand men and absorbed more than 10 per cent of GNP (and half of central government expenditures).[17]

By the beginning of the 1960s, Taiwan and South Korea were, in some respects, in similar economic conditions. Both were heavily dependent on American aid to sustain their socioeconomic structures, had limited export potential as a result of limited domestic investment in export industries and restrictive trade and exchange rate policies, and had low levels of infrastructural and human capital development.[18]

The political-military turbulence that accompanied the emergence of Taiwan and South Korea as modern powers created a legacy of geopolitical insecurity and pronounced dependence for even basic survival on the United States. Over the years, this legacy has fostered an almost desperate aspiration for international status, as well as for greater independence and self-reliance. Dependence on an outside power for survival is almost always a source of national resentment. It also tends to exaggerate perceptions within the dependent country of the potential damage to internal order that a change in policy by the outside power would engender. Each successive 'shock' to the dependent country stemming from changes in the outside power's policy produces the simultaneous reflex of trying to avert future changes – by securing ever more explicitly the commitment of the outside power – while at the same time adopting measures that in the future could minimize the extent to which the dependent country is at the mercy of its benefactor.

Internal sociopolitical and military development in these countries has, to a significant degree, been derived from the threat of internal subversion or military aggression. Martial law, for example, has prevailed in both countries for most of their political histories. This can be explained by their externally promulgated instability. This link between the external threat and domestic political opposition in both countries has resulted in very gradualistic approaches toward political modernization, as defined by traditional criteria of national political participation and the growth of opposition groups.

Political Culture

Regardless of the extent to which political life is circumscribed in the name of national security – and this is the subject of widespread and vigorous debate – political modernization has proceeded apace in both countries.[19] Although it has been a more precipitous occurrence for Taiwan than for Korea, both countries anticipated the need to design internal measures to augment self-reliance in the 1960s, and accelerated these at roughly the same time in the early 1970s. A strong élite is a prerequisite for the type of successful national political integration necessary for the development of a modern society. In rapidly industrializing countries such as Korea and Taiwan, the composition of the élite has to undergo significant transformations over the course of industrial growth to keep pace with the demands of an increasingly complex, technocratic social structure. Even highly centralized governments based on a narrow coalition of interests must rely increasingly on the advice and influence of professionals who have the requisite expertise to advance industrialization and cope with the strains it creates in changing societies. The extents to which this emerging professional class is permitted access to government and is sufficiently enfranchised that it works in harmony with government objectives are significant factors that influence a nation's development path. If a strong consensus is formed among the military, industrial and academic (research) élite as to the desirability of advancing the nation's objectives through defense production, this can serve as a powerful force to unify the goals of the country and to help implement government policy. From a political standpoint, a coincidence of interest among military, industrial and scientific élites can help provide a foundation for a 'development vanguard'. On the other hand, failures to involve emerging or traditional élites in the development process can result in the loss of scientific and technical talent to other countries and, to the extent that disaffected élites remain in the country, cause political instability.

Just as the development of a defense industrial sector – including the scientific and research and development infrastructure needed for it – may coalesce élites into a development vanguard, it may also lead to the excessive transfer of power to an industrial élite bent on using its newly-found influence over the government to advance its own interests rather than the collective interests of the country. Any sector the activities of which are conducted directly or indirectly at the behest of the government will be offered a degree of access to government that would not ordinarily be afforded if it were engaged in pre-dominantly private or commercial activities. In a country where military objectives are paramount, an intersection of interests between military industry (and related research institutes) and the government may afford influence to the industry which extents to decision-making over other important aspects of government planning.

Taiwan and South Korea present two fairly different pictures of the influence which the development of a defense production sector can have on patterns of élite participation in national life.[20]

Taiwan's political development has often been characterized by Western observers as a struggle between ideologues, those who emphasize recovery of the Chinese mainland, and pragmatists, those who stress the importance of internal development and modernization. The interplay between these two fundamental forces summarizes the struggle Taiwanese planners face in seeking to engineer sociopolitical transformation without creating political instability. Defense planning, in particular, requires a strong consensus for its implementation.

The process of industrialization inevitably creates social forces whose rising demands for 'a share of the pie' can threaten political order. To minimize the effects of this factional phenomenon, the ruling élite should be broadly representative of important social and economic groups. Nowhere is this more critical than in the balancing of defense and development priorities to ensure that they are mutually reinforcing. The evidence suggests that in this harmonizing effort, Taiwan has benefitted considerably from the growth of a defense production sector about which various élite interest groups whose support is critical to political stability may coalesce.

Mainland Chinese, who make up 14 per cent of Taiwan's population, occupy most positions in the ruling Kuomintang and have a virtual monopoly over defense decision-making. The native islanders are seriously underrepresented in élite positions.[21] The proportionality of representation of the two groups has changed, however, in recent years in response to deliberate government efforts to accommodate domestic political pressures by including more Taiwanese candidates in local elections and in appointments to relatively senior positions in the executive and legislative branches. Responsibility for all major defense and foreign decisions, however, still resides in the hands of mainlanders and, more specifically, with the president and the few economic and military officials who comprise the Central Standing Committee. There are only nine Taiwanese on this committee, one-third of its total membership. More important, it is generally perceived that those posts of particular sensitivity are not open to native Taiwanese candidates. However, economic opportunities and social mobility for the Taiwanese have increased considerably. Native islanders unquestionably control the commercial sector, accounting for approximately 90 per cent of all export activities. As a result, there is a division in Taiwan between those who have significant control over economic power and those who control political interests, although ultimately the direction of both is shaped through the central policy apparatus controlled by mainlanders.[22]

An examination of the segments of the élite that control government policy on defense production explains in part why the transfer of responsibility for defense industrialization to the private sector has been so slow to emerge in Taiwan. Since the spring of 1980, the Combined Service Forces have been under the control of the president's highly traditional younger brother, General Chiang Wei-kuo. The 'traditionalist' faction of the KMT, including the military leaders who exert significant influence over the disposition of defense resources, is among the most conservative of the mainland groups on Taiwan. This group tends to distrust efforts to widen participation in decision-making, especially in the area of defense, on the grounds that only the mainland factions properly understand the threat to Taiwan from the PRC and can express with sufficient conviction the destiny of Taiwan to recover full control of the mainland.[23]

The present government's effort to involve private firms in defense-related activity could brighten prospects for political integration in Taiwan to the extent that contracts for this work are given to industries owned by native Taiwanese. The interaction between the Taiwanese industrial leadership and government officials in these joint activities could help create habits of and channels for mutual consultation, giving Taiwanese industrialists a larger voice in the defense-planning process. To an extent this is already occurring.

Another segment of Taiwan's élite whose support for defense industrialization has already increased cohesion in national policy and planning are the so-called development technocrats, an aggregate of KMT leaders who exhibit strong convictions about the need for rapid modernization and the continuation of a market economy in Taiwan. Former president Yen Chia-kan is a part of this group. Many Western-trained scientists who have returned to Taiwan to serve in the government and the Nationalist party also belong to it. Some of these individuals have legacies of political exposure as the children of prominent government and party officials. Moderate politically and committed to the goals of political pluralism and modernization, they are a potentially effective force for bringing modern technical expertise and a Westernized perspective on political development to the national policy debate. Pragmatism rather than ideological dogmatism seems to characterize their approach: 'they try to keep a low profile in order not to alienate older leaders, however, they may well be considered as the brain trust of the moderate coalition'.[24]

Although Taiwan has not been completely successful in using defense production to build consensus and instill habits and patterns of co-operation among the island's élites, it has fared better in this than has Korea. South Korea is more homogeneous than Taiwan, having almost no political divisions based on cultural, ethnic or religious lines. The country's political modernization can be described as a product of dynamic tension between the forces of largely Western modernizing values and those of traditional, conservative Korean practices, the latter derived largely from Confucianism. Confucianism provides the ideological underpinnings for popular acceptance of, or at least acquiescence to, the concentration of power in a small, centralized élite with a strong charismatic leader; the élite, in turn, is upheld by a cloesly linked bureaucratic power structure whose conservatism and loyalty to the ruler helps to defend against threats to the status quo.[25]

Although defense industrialization, economic growth and political modernization all made major strides during the 1960s, following the military coup that installed Park Chung Hee as ruler of Korea, it was not until the 1970s that the transformation of greatest interest to this study took place. As in Taiwan, the twin shocks of the American military withdrawal from South-east Asia and the promise of further American retreat incorporated in the Nixon Doctrine stimulated a major acceleration of Korea's efforts to achieve greater self-reliance on defense and other areas.

There is much evidence to suggest that the impact of this period of accelerated change was much more pronounced in Korea than in Taiwan. Although Korea continued to enjoy the benefits of American troops and military assistance in the mid-1970s, the fears which the debacle of American policy in Vietnam inspired among Korean leaders led to intensive policy changes different from those in Taiwan. In the aftermath of bouts of domestic instability in 1971 and fears that alterations in the international environment would further aggravate these problems, President Park proclaimed a state of emergency in 1971 which, translated later into the Yushin Constitution, gave him supreme authority and sole discretion over all external and internal security measures.[26]

National security was the major reason given for this consolidation of national power under Park. Later in 1971, the declaration of a state of emergency contained six points, all of which aimed at reducing public participation in, and criticism of, matters related to security. Thereafter, Park assumed the power to limit freedom of speech and of the press, restrict labor strikes, control wages and prices, and control the armed forces (all without consulting the National Assembly). Internal opposition continued withal. Coupled with continued pressures on Korea arising from changes in US policy and from perceived increases in pressure from the north, the fear of continued domestic opposition led to the declaration of martial law in October 1972.[27]

The legacy of political centralization is crucial to an understanding of the nature of the political élite which emerged subsequently and the way in which it responded to the accelerated drive to promote defense industrialization, for not only did it tend to be reproduced in a highly centralized commercial élite but, because of Korea's cultural homogeneity, the two élites formed a much more cosy and mutually accommodating relationship than was true in Taiwan, where significant cultural differences divided the political and commercial élites. Capping the distinction between the two countries is the fact that the concentration of technical, scientific and administrative skill brought by the 1949 migration from the mainland was not matched by a similarly scaled infusion in Korea.

Beginning from an already considerable technical base, Taiwan approached the modernization of its defense sector by focusing first on the development of scientific capabilities to support it. Divisions between mainlanders and native Taiwanese kept defense research and production largely within government facilities where its progress could be tracked and monitored. The purposeful fostering of interdependency between defense and civilian activity occurred first in the expansion of scientific institutes such as the Chungshan Institute of Science and Technology and the National Science Council (which now coordinates research between industry and the defense establishment). The results have been significant: 'A new pattern for domestic transfer of technology is emerging, and an increasing portion of [research and development] work on components and sub-systems [for weapons is] now being farmed out to civilian institutions.'[28]

In Korea, the line of least resistance for the stultifyingly overcentralized, insular, and technically-underskilled political élite was to cede responsibilities for defense production to 'cousins' (literally or figuratively) in the commercial sector, who received irresistable incentives to undertake the ill-planned and premature production of Korean weapons.[29]

As a result of these policies, an industrial élite came to be included in the small cadre of advisors who had influence over President Park. The political structure that had emerged after 1972 was originally dominated by former military officers, many of whom had participated in the coup and were unquestioned in their loyalty to Park. With the advent of participants from the private sector – based on the cluster of leading businessmen who had been called upon by Park to engage in the 'national development' effort – there was a relative decline in the prestige and dominance enjoyed by military officers. The power structure which began to be shaped in this period thus was one in which an industrial élite – prized for its loyalty but also sought out for technical expertise not elsewhere available to the government – had unprecedented influence.[30]

Government Intervention in the Economy

An active government role in the national economy is the rule rather than the exception in the modernizing world (and in the West as well – the question is one of degree). From long-term planning to final allocations, the involvement of the third world state in economic matters is restrained more by the limits of its practical ability to manage the economic life of its citizens than by its will to do so. Indeed, in the vast majority of states, the will to exercise control far exceeds the capability. In this regard, Taiwan constitutes something of an anomaly. While the mechanisms of control over the population existed in good measure – and indeed were and are freely used in other spheres of political life – Taiwan's leaders appear to have exercised restraint in applying them with too much abandon in the commercial and industrial sphere.

Taiwan's approach to the development and protection of new industries was a fairly subtle one, even in the early postwar years. The government employed only selective incentives to private industry to aid them in their development and to encourage private initiative. Among the measures employed was the establishment by the government of technology and investment institutes aimed at industry to provide credit services as well as management and technical training and assistance. Although such institutes are common in developing countries during early industrialization, the important distinction is that Taiwan's institutes were used strictly to encourage, rather than direct or protect, private firms. As Gustav Ranis has described it:

> While the effort to provide such [incentives] is a common pheno-
> menon in developing countries, the concentration on usefulness to
> the private sector in areas of expanding activity compares favorably
> with the 'search for major breakthroughs' approach to technology
> financed by government subsidies frequently encountered elsewhere.
> In general the [Taiwanese] government avoided the splashy 'white
> elephant' route and seemed more interested in enhancing the capacity
> of the infant entrepreneur to stand on his own feet – presumably
> the basic purpose of erecting the primary import-institution hothouse
> in the first place – rather than to force-feed him continuously into

the need for permanent protection from foreign and domestic competition.[31]

The important difference in Korea during the phase of import substitution in the 1950s lay in the degree of the government's control over the disposition of resources and in its interference with market forces. Under Syngham Rhee, the traditional 'centralization' of political decision-making amounted in practice to despotic and arbitrary rule. Allocations were granted largely according to the principle of cronyism. To the extent that 'infant' industries emerged as the basis for a subsequent industrial infrastructure, they too often depended for their survival on the government's willingness to pass protectionist measures designed to do exactly what Ranis counsels against – foster weak institutions requiring permanent protection from domestic and international competition simply to survive. Corruption aside, the entire industrial base in the early years was simply a dependency of US assistance. More than three-quarters of domestic investment between the years 1956 and 1963 was financed by foreign capital.[32] Although the economy in Korea has since been transformed, with a fairly healthy industrial sector, the weaknesses of the Korean market structure persist, partly as a result of the centralized Korean 'style' of decision-making.

Agricultural Development

Due in part to its colonial legacy, in part to wise government policy, Taiwan's agriculture has weathered the storms of modernization – even the years of intensive import substitution between 1954 and 1960 – relatively well. Sadly, neglect of agriculture is a common characteristic of early industrialization in many developing countries, and results in a severe urban-rural imbalance. This did not occur in Taiwan, largely because industrial policy was accompanied simultaneously by a series of measures to modernize the rural sector, including land reform, extension of credits to farmers, and provision of intermediate technology for more efficient agricultural production.[33]

Among other things, a healthy agricultural sector aids in reducing dependence on imports of food (saving foreign exchange needed for industrial growth), maintains a market for the consumer goods the new industries are producing, and prevents excessive migration to industrial centers with attendant unemployment. The Taiwanese government went further towards rural modernization by adding to its inherited rural institutional and physical infrastructure through improvements in rural transportation and communications, and in education and trade organizations for farmers.[34]

In contrast, rural development in Korea foundered early on, despite American-assisted projects to initiate rural modernization, including, among other things, land reform and the establishment of institutions for providing farm technology to producers. The failures resulted in urban–rural imbalances and an external dependency for food that remains a problem today. The legacy of the Korean War, during which domestic food production declined substantially, was the beginning of a fairly large-scale dependence on outside sources for grain and other primary commodities. The practice of importing foreign agricultural commodities led to a depression in local agricultural prices and aggravated the existing imbalances. After 1961, when the Park régime assumed office, this 'unbalanced growth' symptom became even more pronounced, although officially the government continued to pledge its commitment to agricultural development. In a direct sense, Korea's 'industrialization first, agricultural development second' is quite distinct from the Taiwanese approach, with important implications for the future pattern of growth in both countries.[35]

EXPORT-LED ECONOMIC GROWTH: FUNDRAISING FOR INDEPENDENT MILITARY PRODUCTION

After 1960, industrial production for export came to characterize the economic life of both Taiwan and South Korea. The shift from import substitution to a strategy of export-led growth was conscious and explicit in both countries. Some reforms to the overprotected Korean economy had been introduced prior to 1960, but with minimal effect, and it was not really until the military régime came to power in 1961 that successful export promotion policies, with attendent changes in economic and political structures, emerged. Like Korea, Taiwan even prior to the 1960s instituted a series of economic policy changes, ranging from liberalization of import controls to the adoption of exchange rates favorable to the export trade.

These policies took firm hold during the 1960s. The average annual growth rate of manufacturing production was 17.5 per cent between 1962 and 1972 in Korea, and 20 per cent between 1963 and 1973 in Taiwan. Similarly, Korea's exports rose from $42 million in 1962 to $1.17 billion in 1972, while in Taiwan, exports rose from $174 million in 1962 to $3.34 billion in 1972. As an indication of structural change in the two economies, exports as a share of GNP rose from 5 per cent to 20 per cent of GNP between 1962 and 1972 in Korea, and from 11 per cent to 45 per cent between 1960 and 1972 in Taiwan; the share of manufacturing in these exports, moreover, rose for both countries, increasing from about 20 per cent in the early 1960s to over 80 per cent by 1970.[36]

These dramatic increases in economies that had grown sluggish after a decade of primary import substitution can be traced to common monetary and tax policies designed to stimulate exports. Through the establishment of such trade incentives, the countries developed dual economies wherein exports could be manufactured under conditions approximating free trade, while the rest of industry and the economy as a whole continued to operate with a higher degree of central planning and restriction.[37]

Taiwan's development strategy during this period meets the traditional criteria of classical trade theory, which suggests as two preconditions for successful economic growth the existence of independent domestic entrepreneurs capable of adapting to international market forces (after the removal of protectionist barriers) and a sound infrastructure to 'ensure the sustained contribution of a substantially dynamic agricultural sector'. In addition, Taiwan followed closely the development path embodied in the theory of comparative advantage, by relying to the fullest extent on its major asset – cheap labor – to create competitive trade industries. Labor-intensive consumer goods (apparel and plywood) were traded with the West, where they had a clear cost advantage, while intermediate and capital goods were traded with neighbouring developing countries, with which Taiwan enjoyed a comparative advantage in skilled production.[38]

The growth in exports that began in the 1960s stemmed mostly from production of textiles, clothing, shoes, radios, televisions, electronic components and light electrical machinery. All of these activities are highly labor-intensive, and their production took advantage of Taiwan's low-cost, abundant and diligent work force. Rather than importing expensive (and inappropriate) capital-intensive technology for the development of industry, Taiwan deliberately adapted obsolete and often second-hand machinery, from Japan and elsewhere, in part to save foreign exchange but also, in conjunction with policies to inhibit overly rapid mechanization and to make use of indigenous labor (through multiple shifts and long hours), to maintain the labor-intensity of the emerging manufacturing industry.

Korea's development policies did not deviate drastically from

Taiwan's. Both countries adopted similar measures aimed at achieving rapid economic growth through the expansion of export trade. What is different is the degree of emphasis received by certain sectors, and, as stated earlier, the degree of government intervention in the economy, or reliance on foreign aid or borrowing, and of attention to agricultural development.

The growth of Korea's manufactured exports began from a level that was much lower than Taiwan's: $9.4 million in 1961 compared to Taiwan's $71 million that same year. Exports were concentrated in simple manufactured products such as plywood and textiles. Moreover, Korea did not develop an export capability in even slightly more complex export production – such as shoes or light electronics – until some years later. The rate of progress, however, at least as measured statistically, is unprecedented. Korea's manufactured exports grew at an average annual rate of over 50 per cent. By 1972, manufactured goods represented more than 80 per cent of Korea's total exports.[39]

Among the measures that contributed to Korea's high rate of growth during this period were exemptions from taxes for exports, concessionary loans for export credits, and exemptions from duties of intermediate imports needed for export industries. However, during the 1960s the government created a large web of protectionist measures for export industries. It is estimated that subsidies to encourage production for export amounted to over one-third of the total value of exports by 1970, depleting national government revenues and increasing inflationary pressures. This occurred in spite of a massive currency devaluation initiated in 1964, which had greatly augmented the level of export revenues.

Transcending Geopolitical Place

The Taiwanese economy as a whole manifested its ability to become self-sustaining after 1965, when US aid was terminated. Although Taiwan had been especially dependent on foreign aid between 1952 and 1962, external dependence on grants and concessional loans became negligible after 1965. Foreign capital inflows from commercial and multinational lending quickly took the place of aid. Taiwan encouraged foreign investment through a variety of tax incentives and other related measures, but at the base of Taiwan's attractiveness to foreign investors were its labor policies (low wages and no strikes), infrastructural advantages (transportation and cheap power sources) and overall political stability.[40]

There are some important economic disadvantages that arise from extensive dependence on foreign investment, relating mostly to the absence of linkages from the export sector to the rest of the economy. These did affect Taiwan during this period. Devoted to the production of goods for the export market in areas where the country was cost-competitive, production activity became concentrated in low-to-intermediate manufacturing, a majority of which was in non-durable consumer goods. The spin-offs to the economy in the form of technological advances were limited. This can change, however, as the industrial base shifts toward more technology- and capital-intensive production and a more sophisticated array of goods.

The normalization of Korea's relations with Japan, greatly increased foreign investment, and economic and technical spin-offs from participation in the American war effort in Vietnam all contributed to Korea's successful economic growth during the 1960s. In addition, Korea benefitted significantly from the technical and managerial assistance provided by multinational lending institutions. The International Monetary Fund, for example, played a central role in designing and sustaining a series of programs for stabilizing Korea's economy, beginning in 1965.[41]

Korea has always been more dependent than Taiwan on outside financial support (grants and concessional loans) for its economic development, and Japan played a key role in this regard after 1965. Japanese aid amounted to over $1 billion between 1960 and 1972. Over time, Japanese investment also became a major source of growth and revenues for Korea, as Japanese firms set up plants in Korea to supply both the Japanese domestic market and the United States. This dependence on the Japanese, so soon after Japan's harsh colonial domination, has posed serious political problems.[42]

The US contribution to Korean development has been even more significant. It has included not only large amounts of grants, aid and concessional loans, but also financial transfers arising from Korea's role in the Vietnam War. Earnings from Vietnam made a very significant contribution to Korea's economic growth between 1966 and 1972. The US Agency for International Development estimated that Korean foreign exchange earnings from activities in Vietnam amounted to more than $925 million over this period.[43]

In addition to the transfer of troops to Vietnam (which grew to 47,000 men in the later 1960s and totalled 300,000 between 1965 and

1972), for which Korea was compensated by United States, the country derived revenues from ambitious construction programs in Vietnam. Korea's now thriving domestic construction industry received its first boost from its venture in South Vietnam, where Korean firms received favored treatment and protection from competition in bidding. Overseas construction projects provide a large portion of Korea's national income today.[44]

External indebtedness became an increasing problem over the course of the 1960s, reaching a record high of $1 billion by 1970. Foreign debt of this size not only creates inflationary pressures from deficit financing, but requires large allocations for debt servicing. The gap between domestic savings and investment continued to widen (in part from inattention to the development of background sectors, like agriculture, to serve as a source of savings). The financial costs of government subsidies and incentives for the continued growth of manufacturing exports also continued to increase.[45]

Political Culture

The marked difference between Taiwan's and Korea's political cultures continued to cause divergence in the two countries' development paths. The professionalism and experience of the ruling élite in Taiwan, coupled with the relative political stability of the island, provided the requisite foundations for implementing carefully designed policy. By contrast, the political and social chaos that prevailed in Korea under the tutelage of a corrupt despot suggested that such policy had neither source nor audience for its design and implementation. This changed at the beginning of the 1960s. The élite that came to power in 1961 in Korea did have a technical and managerial orientation, based on military training, but, given the new régime's preeminent need to consolidate its power, its economic plans necessarily reflected a combination of political and economic considerations. The series of economic plans initiated in this decade were often as much a statement of political intent, designed to gain support from key sectors, as they were actual blueprints for economic development. The military officers who had engineered the coup in 1961 entered the public and private sectors in positions of influence after Park assumed office in 1963, helping to ensure that Korea's economic development would be tightly bound up with military preparedness. The result of this direct linkage was substantial investment in areas such as petroleum, chemicals and transport infrastructure, with little regard to the principle of comparative advantage.[46]

Overall, the Korean development strategy during this period was to achieve rapid rates of growth in industry and exports, apparently almost at any cost. The Korean government came to characterize this objective of rapid growth as the ultimate national purpose. As one analyst described it, 'the shift in priorities to a clear economic goal identified economic development with deep-seated national aspirations and linked individual economic ambitions with patriotic aims'.[47]

Government Intervention: the Emerging Contrast

The role of Taiwan's government during the period of export growth was varied and extensive, but not intrusive. Aside from the tax and other incentives referred to previously, Taiwan in 1965 began to create export processing zones aimed at maximizing export-oriented industrial production.[48]

The principal philosophy of Taiwan's economic planners is best described as a mix of free enterprise and free trade policy with a selective application of government incentives and disincentives to stimulate desired production without interfering excessively in the autonomy and competitiveness of private firms. The government's policy of relative non-intervention in the private sector and its emphasis upon independent enterprise in industry was manifested in several ways. For example, selected firms outside the export zones might receive deferrals of import duties imposed on plant equipment but were not exempted from such duties. Throughout the 1960s, the great majority of Taiwanese enterprises were private concerns, with only about a dozen large industrial corporations owned and operated by the state in the areas of industrial chemicals and fertilizers, non-ferrous metals and steel, machinery, and petroleum. The guiding criterion was that new enterprises would be operated by the public sector only if no private entrepreneur could be induced to assume responsibility for them and if these production activities were designated by government as crucial for development. The number of public enterprises increased in the 1970s as heavy and technology-intensive production rose, partly in conjunction with plans for the production of defense-related goods.[49]

Reforms to the Korean educational system in the 1950s paid off by the beginning of the 1960s in the form of a more highly skilled labor force and a small cadre of experienced business entrepreneurs.

Unfortunately, the potential dynamism of these new human resources was greatly attenuated by excessive government protection of industry. In return for abiding by government strictures concerning all aspects of production, private firms obtained government support and subsidies to insure them from risk and competition. The very close 'partnership' that developed between government and industry at this time was based on a very small number of firms. The government owned and operated public service industries (including transportation, power and communications) as well as industries requiring major capital investments, such as steel. Once these firms became established, the government often transferred them to private industry, to those 'favored, carefully selected' private firms the government had designated as rightful recipients and which, as a result, received extensive protection and financial inducements. This pattern of government transfer of industrial resources continues today.[50]

The foundations of Korea's industrial structure are mirrored in the structure of Korea's defense industry. Although actual production of weapons did not begin until the very end of the decade of the 1960s (aside from one or two government arsenals for the production of simple munitions), the economy as a whole was moving in the direction of establishing a heavy industrial base of some relevance to defense. Less attention was paid to developing scientific and research capabilities, however, which became an impediment in later years. As a general strategy, Park perceived rapid industrial growth as the ultimate means by which South Korea could dominate its northern kin.

Taiwan's pattern of industrialization during the 1960s did not emphasize specifically the development of production capabilities related to defense. The Taiwanese labor force did gain experience of direct relevance to later defense production activity in the course of such projects as overhauling US F-4s during the Vietnam War, but there was no productive sector of the economy devoted to defense until the next decade. Particularly lacking was attention to high-precision industry, for example, which would be needed to sustain defense production and develop advanced defense technologies. Nor did related industries, such as machinery and machine tools, grow at this time. More in evidence was a strategy for developing a technical infrastructure for economic growth without engendering serious economic or social dislocations. These are understandably critical objectives for a country whose defense burden had already absorbed over 60 per cent of the small national budget for some years, and in which internal political cohesion and stability were paramount. Implicit in this strategy was a recognition of the strength of Taiwan's economic competitiveness as 'her first line of defense', a belief among planners that persists, in varying degrees, today.[51]

At the same time, there is evidence of a conscious understanding that once Taiwan developed its economic capabilities it would also be able to undertake more ambitious, independent defense efforts. The Chungshan Institute for defense research and development, for example, was established in the mid-1960s, on a modest scale, to lay the foundations for development of defense technology. Infrastructural development, moreover, such as roads, ports and railroads, which received emphasis as a basis for increasing the cohesion of the country and maximizing mobility, is also a necessity for military preparedness, and this could not have been lost on Taiwan's planners.[52]

Agriculture: the Neglected Necessity

Taiwan's strategy of developing a solid rural infrastructure even as industry was developing was further enhanced by the deliberate location of industrial establishments outside the confines of major urban areas. The proportion of manufacturing activity in the major cities was only 34 per cent in 1971, virtually the same as it had been in 1951, while the proportion of manufacturing personnel employed in major cities *declined* from the mid-1950s to the mid-1960s. This distribution of industry helped to maintain stable wage rates, to increase equity in income distribution, and to stem the threat of excessive urbanization, the direct social cost of an imbalanced development pattern. Moreover, it helped to disseminate the developmental benefits of industry – training, education and income – to a larger proportion of the population.[53]

Overall, Korea's agricultural development during this period was not impressive. When the Park régime began its first five-year plan in 1962, it made explicit the government's overriding concern for strengthening the industrial sector. In spite of political statements to the contrary, agriculture was not an important priority. Of the 17 per cent of investment resources committed to agriculture under the plan, the sector actually received only 9 per cent. This allocation declined even further in the second five-year plan, in which agriculture was allocated just over 6 per cent of investment.[54]

This anti-agricultural bias in policy resulted in chronic dependence on outside sources for certain foods and grains that could not be produced domestically. This did not place immediate pressure on the balance of payments, nor tax domestic resources during the period, because Korea was still receiving sizable amounts of US aid under the PL 480 program. This changed in the 1970s, however, and the United States now requires cash or credit sales in US dollars for agricultural transactions.

The neglect of agriculture also resulted in serious urban–rural income disparities in the 1960s. These persist today. In 1962–4, per capita income in the rural sector was actually higher than in the urban sector. It declined to 50–60 per cent of urban wages in the 1967–70 period, when the two five-year plans had taken full effect. As a result, the farm population has been declining steadily. Migration to the cities is estimated to have averaged about 500,000 per year, with all the expected urban poverty and stress on the urban infrastructure, as well as more serious consequences for Korea's long-term agricultural potential.[55]

INVESTING IN SECURITY: THE EMERGENCE OF INDIGENOUS MILITARY PRODUCTION

When the powerful new economies of Taiwan and South Korea were harnessed for the purpose of military production in the early 1970s, the divergent development styles of the two countries stood forth in perfect clarity. This harnessing, while in preparation for a decade or more, was stimulated by significant changes in the geopolitical context within which the two states were obliged to operate. We shall see how Taiwan's restrained and controlled response to the need for a second phase of import substitution – this time including military goods – benefitted from the island's reserves of technical and administrative skill, and, in accordance with the principle of comparative advantage, did *not* include an effort to create a new export market out of the newly substituted goods . . . at least not for the present.

After 1971, both Taiwan and South Korea had to adapt to the political shock of the Nixon Doctrine, which pledged a shrinking of the American military presence in Asia. For Taiwan, this shock was compounded by the US rapprochement with the People's Republic of China, culminating in the Shanghai communiqué of 1972, and ultimately in the 'derecognition' of Taiwan. Taiwan's relations with many former allies and trading partners were transformed irrevocably as a result. These changes would have been sufficient to cause alarm in themselves, but in short order came a series of new shocks to both countries' economies: a rising trend of protectionism in developed countries against manufactured exports from developing regions; inflation in the industrial countries, causing the prices of needed commodity imports to increase rapidly; the sudden increases in oil prices of 1973 and 1974; and a general recession in the developed world, which cut demand for the exorts of Taiwan and South Korea.

Quite apart from these crises, the two countries were facing severe challenges to their export-led economic strategies in the mid-1970s, requiring major economic realignments. This challenge reflected the relative saturation of world markets with the labor-intensive commodities upon which the rapid growth of the 1960s had relied. With rising labor costs in Taiwan and South Korea, a number of new developing countries were threatening to become even more competitive in the world market for labor-intensive goods. Not the least of these was the People's Republic of China. This pointed to the need to restructure the industrial sectors of both states toward more sophisticated products. In the export area, this required production of more complex commodities with solid markets. In the domestic sphere, the new realities also called for a renewed pattern of import substitution for certain critical goods for which import prices were becoming prohibitive. As a direct parallel, the realization that the United States could not be relied upon forever as a supplier of military technology (or as a military ally) required a form of import substitution in the defense area. The latter two goals seemed to call for investment in capital-intensive heavy and chemical industry.

The decision to allocate investment resources to the development of defense industries in the early 1970s formed a central part of overall industrial strategy. However, in the process of implementing that strategy, the two countries' development paths diverged widely, reflecting their individual legacies and the distinct perception of each about how to ensure 'security'. We shall see to what extent that image of security is a creature of the accuracy and appropriateness of past strategic choices.

Taiwan's industrial reorientation was revealed over the course of the 1970s in two new economic plans, the 1973–9 six-year economic plan and its 'second phase' begun in 1973. The three critical elements of these plans were as follows.

(a) Upgrading the quality of industrial products. This was a major priority not only because it was required for the expansion of Taiwan's export markets, but because it related directly to defense.

Prior to 1976, no systematic effort had been made to develop high-quality precision industry of the sort needed for military-grade production. A special aspect of this goal was concentration on advanced technologies useful for defense but capable of yielding significant spin-offs to the civilian economy.[56]

(b) Shifting attention to the development of heavy industries such as steel, shipbuilding, petrochemicals and nuclear power development, and to infrastructural facilities such as roads and highways. This goal also reflected the twin objectives of stimulating the economy while providing a base for the defense sector.

(c) Moving from labor-intensive to skill-intensive industry. This was a prerequisite for continued growth, not only because of poor overseas markets for labor-intensive goods, but because of fears of competition from the PRC in the production of such goods.

The major features of this new policy were the 'Ten Major Development Projects', estimated to have cost $6 billion in government allocations between 1973 and 1979. The projects were concentrated in manufacturing, steel, petrochemicals, shipbuilding, nuclear power generation and transport. The subsequent phase was largely an expansion of the same ten projects, for instance, expansion of the steel and petrochemical complex at Kao-Hsiung. But, more importantly, the modification plan called for an even more dramatic shift into high-technology engineering industries. Machinery, for example, grew at an annual rate of 25 per cent after 1976, compared to 11 per cent between 1967 and 1976. This shift was meant to emphasize greater skill-intensity, more sophisticated technological content, higher value-added, and stronger linkage effects, while lowering capital-intensity somewhat from the levels contained in phase one of the plan.[57]

The government will bear the cost of the capital-intensive industries. It is understood that these will produce not for the export market (Taiwan does not have a comparative advantage in these areas) but for domestic demand, and that they are aimed at strengthening the country's technological infrastructure as well as securing a stable domestic supply of strategic inputs.[58] Given the small size of the market, Taiwan will not be able to achieve economies of scale in this type of production, however, and this will raise costs. The burden on the government budget, finally, may require diversion of resources from other types of investment.

The apparent diseconomies of capital-intensive production that Taiwan is willing to incur at this point can be partially explained in terms of defense considerations. In the area of nuclear and geothermal energy, for instance, the development of power plants and increased exploration are obviously aimed at reducing dependence on potentially unreliable and increasingly expensive sources of oil.[59] Steel production can also be explained as a direct corollary of the defense industrialization effort, as it reduces uncertainties in supply. The extent to which these projects contribute to or interfere with Taiwan's economic growth, however, will determine whether the current modernization strategy is advancing security (in military and economic terms) or opening the island's economy to new threats.

In reviewing Taiwan's development strategy during this period, it is clear that the country has generally avoided excessive investment in inappropriate capital-intensive industries that would overtax the economy and inhibit growth. In view of severe shortages of energy and manpower, added to the heavy demands of national defense in the years since derecognition, Taiwan may opt to discontinue some of its major development projects. Fortunately, the government has not staked its political prestige on the continuation of any of the projects.

Taiwan's incrementalism and selectivity in moving toward capital-intensive development stand in sharp contrast to the measures adopted by Korea. Like Taiwan, Korea embarked in 1972 upon an ambitious plan to improve the quality of industrial technology and pave the way for an economic transition from labor-intensive light industry to skill- and technology-intensive industry. The announcement of planned US troop withdrawals under the Nixon and Carter administrations coincided with a two-phase acceleration of heavy industry investments under the Park régime to provide foundations for rapid development of defense industries.

The Park régime was also inspired in large measure by what it considered to be its previous successes in industrialization during the 1960s.[60] Building on the cooperative relations established between the small cluster of businesses and the government, and relying on deficit financing, the Park régime managed to increase gross domestic capital formation by 27 per cent in 1977 and 30 per cent in 1978. Most of this was in heavy and chemical industries or in projects that directly supported these industries.[61] By all accounts, including those of Korean economists, expansion of these industries greatly exceeded the production levels that would have been consistent with the size of Korea's market, financing capacity, and technical and engineering abilities. The concentration of projects in a few large firms, all of which stressed self-sufficiency within the firm to maximize profits, led to extremely high levels of duplication in capital-intensive facilities. Economic benefits were not disseminated widely since there was no development of supporting industries to supply the large, vertically integrated enterprises.[62]

The haste with which heavy industries were developed was replicated in the defense sector. Defense industries pose a particular problem for the Korean economy in that they are even more heavily subsidized by the government than are their civilian counterparts, but do not directly produce significant positive growth. Heavy subsidization is a cost that industrial governments have traditionally borne as the necessary cost of maintaining independence in defense planning. In Korea, however, this burden was not only excessive relative to the normal efficiency of such production, but was out of proportion to the contribution of the defense firms to security. Production of small munitions and ammunition, for example, the bulk of Korean defense production to date, far exceeds domestic demand and has only limited export potential. It is also an area of production in which there are only limited spin-offs for the civilian sector and for technical advancement, with the exception of the training of semi-skilled workers in routine operations. More advanced defense production projects, such as the F-5 program, do produce spin-offs, but their costs are much higher.[63]

The widespread phenomenon in Korea of overinvestment and misallocation in heavy industries (and the resulting skewed pattern of industrial growth) places into question both the strength of the private enterprise system in Korea and the sufficiency of official institutional arrangements to plan these investments.

The cadre of private entrepreneurs who developed close relations with the Park régime and became a *de facto* part of the planning apparatus in the 1960s tended to be the heads of large business conglomerates. Many of these firms had entered into industrial activity on the basis of government loans financed by external indebtedness. As early as 1969, the Korean reflex to channel money into private firms that had no demonstrable ability to produce efficiently had created serious economic problems for the government. In August of that year, for instance, the government had been forced to assume responsibility for over thirty firms that could not meet their debts. By 1971, almost two hundred firms had gone into bankruptcy.[64]

The firms that survived to become the favored enterprises of the 1970s remained highly dependent on government inducements and preferred loan treatment. Nowhere was this more true than in the defense sector, where firms were guaranteed 10 per cent annual profits by the government and received a wide range of tax and other financial inducements to enter into defense production.

The political and economic benefits of defense industrialization were enjoyed by such a small number of Korean firms that other companies in related activities were unable to compete successfully, further reinforcing the tendency toward government-dependent monopolies. Moreover, there were sufficient instances of graft and conspicuous consumption by the newly-enfranchised industrial élite to invite political controversy and challenge the legitimacy of government practices.[65]

Preferential treatment toward favored heavy-industrial producers took the form of allocations of scarce investment funds at the expense of medium and light industry, with the latter consequently facing severe difficulties in raising necessary capital for production. Since the government controlled credit, it simply ignored market forces and determined decisions about investment according to its own priorities. Investors received government subsidies and other inducements, and did not coordinate production plans with one another, resulting in duplication of capacity and outright inefficiency.[66] As an example, a major area of inefficiency – arising in large measure from government policy – can be found in the machinery sector, a sector of considerable significance to Korea's long-term defense modernization objectives.[67] By early 1980, this sector, which had received over $1.4 billion in investment resources since the mid-1970s, was suffering from serious overcapacity. Exports suffered because of poor quality, technical manpower was insufficient to sustain plant equipment units, and a number of firms were near bankruptcy. In heavy electrical equipment, the leading producer was operating at below 50 per cent capacity, while in marine diesel engines capacity utilization was below 30 per cent.

The government's response in late 1980 was to attempt to consolidate this sector in a fashion that directly parallels the measures currently being considered for the rescue of some defense industries. Included in these measures was the selection of a prime contractor for plant equipment, Hyundai International – which was designated sole supplier of such equipment – under a new corporate structure for which Daewoo Heavy Industries was to have management responsibility. All other producers in this sector were prohibited from serving the domestic market. The point was to eliminate some duplication while still permitting the largest firms a share of the market. Renamed the Korean Heavy Industry Company, the firm was quickly subject to expansionary efforts by Daewoo, which wanted to add plant construction to its activities. However, Daewoo ran out of funds to finance such projects, and became overextended. The government then made the firm a public corporation, and infused it with $400 million in additional capital.[68]

Under the leadership of Chun Doo Hwan, the government seems to be making reasonable efforts to rid itself of the responsibility for inefficient firms and is working to undo corruption and government favoritism. Chun's declared priority since taking office in 1981 has been to 'insure clean government' from which 'hidden pockets of corruption must be uprooted'.[69] The consolidation measures for overextended industries are likely to prove more difficult than is currently acknowledged by Korean officials and will require a substantial restructuring of industry, with major economic losses for industrialists. This would result in the fairly rapid disenfranchisement of some industrial leaders who have become quite influential over time. Chun's efforts to revive the importance of market forces and restore competitiveness in industrial development will also require the support of the scientific and technical infrastructure, and of independent entrepreneurs. The two major economic planning units, the Economic Planning Bureau and its research arm, the Korean Development Institute, probably do have the rare talent and expertise needed to plan allocations and project demand in a more rational fashion, but so far have been inhibited for political reasons from doing so.

Although a large number of technically trained personnel became involved in the decision-making process as a result of defense industrialization (and industrialization as a whole), Park's highly centralized governing style prevented the skills of such personnel from ramifying freely through the industrial system. Even as the cadre of decision-makers grew in size and diversity at the top levels of government, individual prestige came increasingly to depend on access to President Park. As Vreeland described it in the early 1970s, 'the extent of political influence in South Korea was measured by physical proximity to the person of President Park. Formal institutional position was not the decisive criterion in the enjoyment of power'.[70]

The officials who occupied places of prominence in the Park régime recommended policy to the president on the basis of decisions taken consensually.[71] The fear of losing Park's regard, however, was so intense that advice on policy was often tailored to avoid bad news. In the area of defense production, this affected defense planning to a considerable extent. Projects that were not particularly successful were often continued, partly in response to pressure from Park himself, partly from fear that reporting failure would result in the disenfranchisement of the agency or of cabinet leaders in charge.

The inefficiencies that arose as a result of the centralization of authority in President Park increased over time as the president became more isolated from his small cadre of advisors. A variety of reasons are given for this progressive isolation, ranging from the 1974 murder of Park's wife (who served until her death as his active liaison with important segments of the public)[72] to the success that Park had enjoyed with projects undertaken against the advice of technical advisors.

Regardless of the reasons, meaningful political participation in Korea, even at the élite level, became progressively more circumscribed under Park. Accompanied by increasing public unrest due in part to Park's repressive policies, intrigues and strife among the president's internal policy élite finally culminated in his assassination in October 1979. In spite of the extraordinary economic successes his régime brought to Korea, Park's failure to broaden the base of political participation in government eventually brought his demise. Rather than forging links among influential sectors on the basis of shared goals – defense industrialization prominent among them – Park's policies created a narrow élite that tended to serve its own interests and over time degenerated into factionalism and instability.

The success of Park's political career had been based in large measure on the support the president enjoyed from the politically neutral armed forces. Deterioration of the domestic political order in South Korea eventually affected military morale. A perception among Korea's military that the existing leadership was corrupt and unprofessional was a major influence. Few in Korea doubt that, for the foreseeable future, the final arbiters of power in domestic affairs will be the armed forces. Nevertheless, the Chun government reflects only a partial reascendance of the military in Korean politics. One lesson of the political and social chaos that prevailed prior to Park's demise was that the Korean economy and sociopolitical structure had grown too complex to be entrusted solely to military management. Chun's choice of advisors reflects this recognition. Although his immediate aides are military men, they are counterbalanced by American-trained economists, for example, Kim Key Hwan, who was made president of the Korean Development Institute, and Kim Jae Ik, member of the presidential staff.

Korea's economy, much more than Taiwan's, is often cited as a case of successful 'trickle-down', or unbalanced-growth, development. While this was true to an extent in the early 1970s, it no longer seems to be accurate. Government control of the financial system and of allocational distribution may have been effective in the past in directing the Korean economy toward rapid growth in industrialization, but as that economy grows more complex, it can no longer be so ordered. As Korea has moved into the 1980s, the critical economic imbalances that have existed to varying degrees in the past – a weak agricultural sector, inefficient industries heavily dependent on government assistance, extensive overseas borrowing to continue artificially high rates of domestic expansion, and strains on the social infrastructure – all have become worse, not better. As a result of success in technological advancement in some sectors, moreover, there are now more pronounced requirements for sophisticated managerial and technical skills to adapt and develop new technologies and production systems. As industrial modernization proceeds, shortages in inputs – skilled labor, energy, domestic savings – will also become more, rather than less, significant for continued economic progress.

Theories of unbalanced growth tend to overestimate the capacity of the developing society to adapt in a healthy, 'organic' way to the changes introduced by and through the favored sector. In Korea, the Park government's subordination of all economic and social goals to the grail of rapid industrial growth has strained Korean society in many ways.

The pace of urbanization in Korea was one of the fastest in the world in the late 1970s, and shows no signs of abating. In spite of the development of urban centers in the southern region, partly owing to deliberate efforts of the Park régime to open this region to industry and tourism, the population remains highly concentrated in the two cities of Seoul and Pusan. Between 1970 and 1978, 50 per cent of Korea's population growth occurred in Seoul. The increase in urban density has resulted in a shortage of housing that could quickly become a source of political instability.[73]

In addition to underallocation for housing construction, there has been relatively little expansion of educational facilities, even though the number of school-age persons has increased. Overcrowding hurts the quality of education, a serious problem in a nation facing severe shortages of skilled manpower.

The Park régime had ambitious plans for curing the ills of excessive urbanization. Beginning in April 1979, the construction of new industrial facilities was banned in Seoul itself and in the northern part of the Seoul metropolitan region. A new industrial city was begun in

Banweol, near Seoul, to accommodate new industry that was to be relocated by government decree. Incentives were provided to firms locating in target regions, including Daejon, Gwang-ju, Jeonju, Masan and Doegu, with industrial estates built near each city. The cost of forcing or inducing firms to relocate to new industrial towns is obviously very high. As the plans have progressed, moreover, towns have developed around industrial complexes. As the firms in these complexes (most of which are heavy industries) move through recessionary cycles, entire regions face the threat of economic dislocation. This problem plagues the industrial West as well. It has been faced, more or less wrenchingly, in the mill towns of England and New England, and in the steel towns of France and Pennsylvania.

The problems of inflation and energy dependence have been equally troublesome for the Koreans. Korea had double-digit inflation throughout the 1970s, an effect due increasingly to domestic rather than international forces. Prices for housing, consumer goods and food escalated rapidly from 1975 to 1980, as demand for goods and services rose with per capita income. There were serious shortages of consumer goods and industrial inputs, the direct result of sectoral imbalances. Another, even more important, cause of inflation was rapidly rising wages in industry. In anticipation of rising prices, workers in the late 1970s demanded increases in pay. As a political expedient, a characteristic 'quick fix', the government permitted the 'informal indexing' of wage increases to inflation, with predictable effects on prices.

A major impetus to inflation has been the cost of energy. Korea by nature has high energy requirements, but even so, given the structure of production in energy-consuming activities, the demand for oil has escalated sharply over the last decade. Paradoxically, it was shortly after the oil crisis in the mid-1970s that Korea redoubled its efforts to develop energy-intensive industries. Between 1979 and early 1980, Korean planners raised domestic oil and power prices dramatically in an effort to pass on to consumers the full cost of imports and services. Oil prices were raised by 75 per cent in 1979 and another 60 per cent in 1980. The effort was aimed at reducing consumption, but it also caused prices to climb throughout industry.

Energy diversification and conservation are extremely expensive and have been slow to take hold. For instance, nuclear plants and hydroelectric projects have required large government investments. Apart from financial constraints, there is the more serious shortage of skilled manpower to design and construct production units for alternative energy sources. This deficiency is beginning to be rectified through the development of research institutes the express purpose of which is to acquire knowledge about new technologies, but this will take some time.[74]

Inflation has several effects on economic stability and development, but the most important for a trade-dependent country like Korea is the threat to the long-term competitiveness of exports. Real export growth declined precipitously between 1976 and 1979 (36 per cent in 1976, 19 per cent in 1977 and 14 per cent in 1978), reflecting wage increases not accompanied by commensurate increases in productivity.[75]

Compared to Taiwan, certainly, Korea did not benefit as fully as it might have from the development of a defense industrial sector. Preferring for the most part loyalty over expertise, the government structure has not made a smooth transition from its traditionalist past to modern administration. An overly close relationship between government and business, not mediated as in Taiwan by a certain cultural distance between the two groups or by the presence of a strong, independent and indispensable scientific-technical-academic élite, deprived the country of a source of necessary and salutary tension.

As a last word, because it can enable a society to absorb the shocks of international market fluctuations and limit the political side-effects of unbalanced growth at home, the agricultural sector deserves one last mention. The Korean government attempted to boost farm production and incomes during the 1970s by pursuing programs such as price supports for key products (rice and barley), by providing low-cost fertilizers, and by restricting agricultural imports. High-yield seeds also helped to boost rice yields. Production did improve between 1968 and 1973, but ultimately the programs responsible for the boost engendered more problems than had existed prior to the intervention. For example, the scheme developed to uphold prices – consisting of government purchases of specified quantities of grains – resulted in budgetary shortfalls in the administering agency as the government failed to pass on the higher prices to urban consumers. A scheme to provide low-cost fertilizers through a government-supported fund produced similar results. In 1980, deficits from both operations amounted to fully 8 per cent of government revenues, or 1.4 per cent of GNP.

5   Conclusion: a Broad View of Security

Many aspects of the postwar economic development of Taiwan and South Korea can be persuasively explained as responses to US foreign policy. In some respects, US policy toward the two countries has been similar. In others, it has been very different, and the results have differed accordingly. Of course, the unique political culture and colonial history of each state has also shaped its responses to US moves and helps explain why, under similar circumstances, different choices have been made.

Viewed in perspective, however, it is hard to see certain prominent features of these two small 'garrison states' – including the vigorously pursued development of their indigenous military industries – as anything other than adaptations to the changing international environment. The single most important fact of that environment for both states has been their intimate relationship with the United States.

Since these relations were established, the United States has sought intermittently to rid itself of unwanted obligations while its partners have struggled to maintain certain vital manifestations of support. At the same time, the United States has, often vainly, endeavored to retain full control of its once totally helpless creations, while the latter have grown, felt the thrill of power, and chafed under the sometimes suffocating mantle of their patron. The tension inherent in any such relationship is an engine for change. As actions and counteractions multiply, accompanied by misunderstandings and misperceptions of varying intensity and consequence – and by the occasional happy entente – relations gradually evolve. At some point, observers agree that a new order has replaced the old. When this point is reached in relations between the United States and the nations on its defense perimeter, the growth of indigenous military industry within these nations is certain to be singled out as the 'cause' of the change.

In this concluding chapter, we shall focus on defense production as both cause and effect. A result, clearly, of US efforts to cast off peripheral responsibilities as it dealt with the 'failure of containment' in Vietnam, military industrialization in Taiwan and South Korea has also had decisive effects, some fully anticipated, others less powerful than hoped, still others very surprising to the United States and its partners alike.

It is useful to reintroduce here the deceptively simple concept of self-reliance. As with 'security', 'dependency', 'modernization', 'development', 'stability' and other staples of the international relationist's vocabulary, self-reliance hides at least two basic meanings. The Nixon Doctrine of 1969 was presented as a policy that advocated self-reliance. By it, the United States intended Korea and Taiwan, among other allies, to bear more of the brunt of the common defense. The allies naturally enough viewed self-reliance in more nationalistic terms. And it was as an expression of nationalism that the concept took root and prospered. It would be well to keep both senses of the term in mind.[1]

Our study has shown that through the growth of military industry, the goal of self-reliance in defense has created linkages within the domestic economies of Taiwan and South Korea, and between these economies and the international economic and military system. Many of these, notably the effects of defense industrialization on human resource development, have had lasting, beneficial effects on the overall development efforts of the two countries. The responsibilities of planning, financing and executing military projects at home have also led Taiwan, and to a much lesser extent Korea, to a more realistic appraisal of the threats they face, and to more temperate strategies for meeting these threats. In a very general way, then, self-reliance has, in quite different degrees, contributed to the healthy, versatile economy and reasoned defense policy that, with the help of just enough well-oiled hardware, make for security. Unfortunately, the heady quest for self-reliance has also brought disappointments. After beating the drum of national independence for two decades, Taiwan and South Korea are still today forced to acknowledge that dependency remains: dependency for loans (in Korea's case), for re-export licenses, for vital components, and for complete systems that cannot be produced domestically. The United States has suffered disappointment, too. Using their growing engineering abilities, the Taiwanese and Koreans find means of acquiring embargoed components for restricted systems. Both states persist in flouting US wishes (and the terms of licensing agreements) in re-exporting military hardware to third-party states. Finally, both states have been self-reliant enough to produce weapons capable of offensive and very destabilizing use, but not nearly enough to ensure the 'common defense' without continued US prodding and assistance.

HAVE INDIGENOUS WEAPONS MADE TAIWAN AND SOUTH KOREA MORE SECURE?

Security is . . . a flexible, productive economy that bends with the shifting winds of international trade; a polity that works out its differences within – and without destroying – its political and social institutions, and without looking about reflexively for scapegoats in times of trouble; a geopolitical space that offers some diplomatic maneuvering room; and, finally, a credible deterrent force, the military might to support a nation's diplomatic choices.

Political Security

Change is painful. The massive social changes that accompany economic modernization – as traditional ways give way to factory discipline and the society urbanizes and rationalizes – can strain political institutions built in and for another time. To weather these strains, the population of the changing society must be able to agree on certain powerful developmental goals that make the pain of change seem worthwhile. (Force can stop occasional or even rather frequent seditious and secessionistic eddies, but it cannot substitute entirely or for long for the allegiance and the goodwill of the population.)

The development of military industry, which struck sonorous nationalistic chords among the Taiwanese and Korean people in the 1970s, seemed ideally suited to help mold the political consensus needed to overcome the centrifugal force of rapid economic development. In varying degrees, the defense-industrial sector has indeed helped Taiwan and South Korea to create a shared sense of national purpose among important segments of the population. At the same time, however, both countries have experienced types of political instability that are widely attributed to the failure of the ruling élite to accommodate the growing demands of a modern society. Korea's political instability has been more marked, of course, but in a sense both countries share the same problem in finding a durable national *modus vivendi* that will protect against internal subversion without requiring the types of political controls that bring on the very forms of political unrest they are designed to eliminate. The pervasive sense of threat that prevails in both countries explains in large part why the government has been able to avoid or delay offering the gradual liberalization of political life newly-enfranchised economic groups come to expect. All but the most radical of opposition groups want to avoid actions that might embolden foreign opponents seeking to take advantage of prolonged political turmoil.

In Taiwan, the first fact of political life is the division between the mainland minority and the Taiwanese majority. This cleavage is partially reproduced among the élite in the separation of ideologues and pragmatists. These splits tend to be less pronounced when it comes to issues of national defense, although serious differences over points of military strategy and policy certainly exist. The expansion of business opportunities through defense contracts has reinforced links between the Kuomintang government and the Taiwanese-dominated commercial sector, and has done so in an area of central political importance. The care with which economic planners on Taiwan have implemented economic growth policies to ensure relative equity in the distribution of income and the other benefits of growth has helped to temper potential sources of political dissent that could otherwise have arisen from the maldistribution of political power. Occasional outbreaks of discontent aside, economic well-being has, on the whole, broken down many political barriers, enlarging the political fold. (It must be acknowledged, however, that significant segments of Taiwan's intelligentsia remain outside this fold.)

Although it does not appear on the surface that the basic structure and ideology of the ruling régime have changed over the past two decades, there has been a subtle evolution in Taiwanese political life in which the growth of the military industries has played a part. More influence is now accorded to experts, in defense and elsewhere, and there has consequently been a reversal in the brain drain, previously a major impediment to national development. This has occurred at least in part as a result of the growth of the defense-related scientific infrastructure, a product of deliberate government policy to include the country's academic sectors in defense research and development efforts. Given that discontent among students and intellectuals poses the most salient threat to political stability, harnessing their talents and attention in this manner is an important political tactic. Moreover, the industries and research institutes related to defense should also help to overcome the anti-military class bias of the commercially oriented native Taiwanese, thus increasing their readiness to support defense efforts.

The increasing demand for skilled labor and management commensurate with Taiwan's rapid industrial growth has created labor shortages in certain sectors, requiring in some instances the use of foreign labor to fill in the gaps. An effort to rectify labor shortages through the expansion of vocational training in Taiwan is presently increasing the number of individuals available for industrial employment. To a considerable extent, this has helped to break down another class-related bias against alternatives to liberal arts education. The awareness that better avenues to career success are now available to graduates of vocational schools – including jobs in the prestigious defense sector – has contributed to this change. The government has reinforced its policy by providing exemptions from military service for young men assuming positions in enterprises deemed vital to national defense.[2]

Korea's political divisions have been less responsive to industrial growth, civilian or military. Years of political centralization in the decision-making apparatus of the Korean government, accompanied by the growing isolation of President Park, left a legacy of personalistic, as compared to pragmatic, determination of economic policy. The gradual dispersal of political and economic power to what are still highly-concentrated business interests is a step up from the prior centalization of all authority in the Blue House, but because the apparent decentralization of economic power was not accompanied by reductions in government subsidization of industry, it has not brought about a corresponding spread in the willingness of entrepreneurs (or planners) to assume responsibility for decisions made. As one scholar has noted with respect to public management in his country: 'The main task . . . is that of maintaining accustomed procedure, not solving problems. . . . Rights and duties defined in terms of job description are meaningless because the actual power always lies in the hands of some small clique formed around a powerful figure.'[3]

These patterns of administrative behavior are replicated at the corporate level. The failure to anticipate adequately and to solve problems in production as these occur, for instance, has resulted in poor quality of output in a number of projects. From a larger perspective, the problems of overcapacity and general inefficiency in the defense sector arose from the failure of Korean leaders to engage in rigorous planning and to accept objective criteria, rather than perceived prestige, as valid guidelines for industrial development.

These and other impediments that Korean planners have faced in their attempt to adopt more efficient forms of project design and management all seem to stem, to an extent, from Korean traditionalism.

In his Yushin system, President Park attempted to meld Western production patterns with Korean traditionalism, and build social consensus around the goals of national security and economic prosperity. Military industries seemed the perfect embodiment of the Yushin ideal, offering an avenue of close cooperation among economic and military interests. In practice, however, the inefficiencies inherent in the concentrated pursuit of defense industrialization, seemingly as an end in itself pursued without adequate consultation with military planners, led to more, rather than less, friction between the military and the cadre of defense contractors favored by Park. When in military circles the perception became firmly rooted that military objectives were being subordinated to economic favoritism, Park was toppled. The headstrong emphasis on economic growth promoted in the Yushin ideal undermined social values unrelated to growth, as 'economic ideology was substituted for ethical and moral ideologies'.[4]

The efforts of the Chun government to rid Korean society and bureaucracy of corruption and favoritism have been extensive. Accompanied by steps to liberalize the political system, including removal of the nationwide curfew that had been in effect for thirty-six years and the reduction of sentences for a number of political prisoners, Chun's efforts to restore public confidence in government are important steps toward greater political cohesion in Korea. Moreover, in direct contrast to Park's isolation, Chun has made important strides toward greater public access to government, symbolically through public appearances, but also in more sustantive ways.

Chun's popularity with the Korean people may be bolstered by his efforts on behalf of Korean reunification. In 1982, Chun outlined in a major public speech – the first by a head of state since 1968 – a comprehensive plan for the unification of the country, the most detailed and complete program ever offered by South Korea. Since unification remains a major issue among Koreans, this is a significant step, one which Park, who hoped for little in dealing with the North, had been reluctant to take.

Nevertheless, the strains of the Yushin era remain. While thousands of Koreans have found employment in the nation's defense industries, they share the teeming cities of Seoul and Pusan with thousands more rural migrants who responded to the call to 'economic prosperity' only to find unemployment and dire urban poverty. The Chun government's efforts to consolidate defense firms with idle capacity have the potential of displacing even more workers. With skills that may not be easily transferred to civilian industry, the new unemployed will be ready recruits for the sort of militant activity that is likely to occur whenever a prolonged period of rising expectations and gratifications is followed by a period of sharp reversal (during which the gap between expectations and gratifications widens and becomes intolerable).[5] When government is highly interventionist in all areas of economic and social life, as it is in Korea, the frustration that accompanies adverse changes in economic conditions can easily become channeled into unrest directed at the régime.[6]

For the moment, however, the development of defense industries in Korea and in Taiwan is advancing political modernization, slowly in the first case, more quickly in the second, through its broadening effect on the participation of individuals and groups in national institutions which, though essentially industrial, are of central political consequence.

Economic Security

Taiwan's development strategy reflects a fairly balanced mix of defense and civilian modernization objectives intended to harmonize the competing demands of various economic sectors and advance structural change incrementally, thus avoiding many of the social and economic strains associated with overly rapid industrialization. The government has successfully guided industrial growth without excessive intrusion, moreover, permitting the ascendance of a capable industrial class that has founded and administered efficient industries.

Taiwan's planners seem to have perceived that military capabilities, crucial as they are, could not ensure the country's defense in the absence of the international support that the expansion of trade relations made possible. By augmenting the number of countries having a stake in preservation of the island's stability and by providing foreign exchange necessary to continue overall modernization, Taiwan's economic strategy became, in an sense, the island's first line of defense, the major palliative to a diplomatic isolation that otherwise would have proven fatally corrosive to the island's stability and national integrity.

The commercial competitiveness that has made possible Taiwan's continued access to the international system since the early 1970s has made Taiwan's planners wary of economic development schemes that might undermine the growth and stability upon which this commercial access depended. Until recently, therefore, Taiwan has avoided the development of capital-intensive industries and other such investments that could have strained the country's resource base and from which economic returns would not have been immediately realizable. Instead, planners have focused on technology-intensive industries, many of which are directly or indirectly useful for civilian and military production of the sort Taiwan required in order to stay one step ahead of its foremost opponent, the People's Republic of China. Taiwan's planners have long understood that the PRC could wage an 'economic war of attrition' in consumer products to undercut Taiwan's economic competitiveness. Given the lower cost of labor in the PRC, trade war of this kind is a realistic scenario and would constitute a security threat of the first order. The technology-intensive production needed to prevent this economic threat from becoming reality is also the key to Taiwan's qualitative superiority in weapons. More will be said on this subject below.

Korea's case is different from Taiwan's. The connection between Korea's fear of abandonment by the United States and its efforts to make great developmental leaps through rapid industrialization is unmistakable. Korea's leaders responded to security challenges, real and imagined, with ever bolder measures, liberally and uncritically financed by bilateral and multilateral assistance. Unfortunately, the centralized and personalistic fashion in which economic policy was implemented (usually without the aid of reliably objective advice from independent defense planners) did not place enough emphasis on long-term project coordination and evaluation.

Korea's development strategy reflected directly the basic defense priorities of President Park. The aspirations to achieve rapid economic growth along with rapid military modernization were inseparable aspects of Park's vision of achieving superiority over the North. Even as South Korea achieved this superiority, he argued, the economy would grow sufficiently to ensure permanent domination of the peninsula by the South, both locally and in regional and international affairs. In spite of the strains associated with Park's ambition, the basic objective of economic superiority was, in fact, achieved. There are still sufficient weaknesses in South Korea's weapons production capabilities, however, to make superiority in this area less definite.

At the same time, the heavy-handed government control of the economy that has plagued Korean development has stymied the emergence of an entrepreneurial class capable of generating efficient and self-sustaining industrial growth without reliance on government subsidies. Even as industrialists have received special favor, moreover, rural development has been neglected. Rural underdevelopment has added to the country's external dependency, and may be an important source of economic and political disruption if agricultural stagnation continues to drive the rural populaton into overcrowded urban centers. A shortage of jobs for unskilled labor is a related problem that will require careful management if it is to be solved.

The economic dislocations that may be expected to follow from efforts to rationalize the overbuilt defense industry will be high. Economic stability may be further endangered if, during the restructuring period, Korea undergoes more general dislocations of a force sufficient to provoke a major recession. Were political instability to accompany economic disruption, Korea's problems could be compounded by the flight of foreign capital out of the country. Korea's excessive dependence on foreign capital, a factor aggravated by continued increases in defense budgets without regard to domestic financial limitations, is riddled with risks and should be a source of primary concern to the Korean government.

Taiwan is still highly dependent on what remains of its official relationship with the United States, as this continues to enhance the legitimacy of the ruling party. Over time, however, the country has been learning to cope with minimum American assistance, assistance which its leaders recognize is likely to become even more limited in the future. Barring further abrupt ruptures in US–Taiwanese relations and unexpected threats to the island's stability, the gradual ascendance in decision-making of technically-proficient moderates is likely to continue and will have a salutary effect on Taiwan's independent political and military development. Interactions with consultants and other paid experts will engender different patterns of relations between Taiwanese and foreigner than were evident in past relations with the US armed forces. The hiring of consultants suggests a more equal partnership, requiring extensive participation by Taiwanese planners in the choice of consultants, the implementation of studies and projects, and the evaluation of results. If only to protect themselves from unsound advice, Taiwan's leaders will have to rely on indigenous talent working alongside foreign experts. This is clearly different from responding to the dictates of a foreign power, a power which, by the same token, could be held responsible if things went awry.

Diplomatic Security: Bargaining Chips and Sunk Costs

The expansion of military industry in Taiwan and South Korea has enlarged the diplomatic maneuvering room of both states in four fundamental ways. Firstly, possessing a credible capability to engineer and produce a weapon system not customarily available for import greatly improves the likelihood of being able to bargain successfully for purchase of the system. Secondly, dual-use technologies and components for systems that for political reasons are not available from suppliers fully-assembled can often be acquired bit by bit, and thus hidden amongst the high-technology needs of a complex economy. Thirdly, all weapons manufactured domestically, including coproduced weapons, enhance the 'nuisance value' of the state that possesses them. Regardless of the terms of licensing agreements or alliances, weapons in hand have the potential for being sold or used. While no state may repeatedly threaten such sale or use without fear of distasteful consequences, it is obvious that, with the weapons, the state has more diplomatic options than it did without them. Fourthly, coproduction projects decrease the maneuvering room of the supplier state as its interests and the interests of its nationals in the coproduction grow.

Both Taiwan and South Korea have exploited potential abilities to deploy weapons of indigenous origin to pressure the United States into more permissive policy on transfers of advanced American technology. Taiwan, as a matter of policy, carefully chooses among production and import options in a manner made possible partly by the island's implicit threats to develop certain weapons independently in the event of US denials of exports. The notion in the United States that some leverage is maintained over Taiwan's policies through transfers in fact grants that country more latitude in force planning than was the case when no production facilities existed.

Similar tactics have been evident in Korea. Conventional (and nuclear) weapon development programs have served as a mechanism for receiving more advanced technology imports, particularly missile technologies, in return for ceasing development activities in these areas inimical to the United States.[7]

On our second point, the high cost and political difficulty of importing the advanced technologies necessary to sustain up-to-date forces (self-reliance and coproduction notwithstanding) have led Taiwan to begin a process of military 'import substitution' wherein more advanced indigenous weapons – missiles in particular – are fulfilling local requirements. This has been made possible by the availability, through commercial channels, of dual-use technologies and components that, although useful in systems which the US refuses to sell Taiwan, are in themselves not strictly controlled for export by the US government. Although the United States has made some efforts to control the flow of components with particularly sensitive potential military applications (for example in missile development), this type of export channel allows the United States to assist Taiwan without provoking from the PRC the strong objections that the transfer of complete systems would entail. Incremental transfers of technologies and components for weapon production projects can be carried out through channels which often do not require public review, as has also been the case for the logistical transfers for several Taiwanese weapon programs. These latter have proceeded uninterrupted for many years, even during the period of the embargo on sales to Taiwan that began in 1979. Taiwan still needs some high-visibility items – to update its forces and bolster its political-military status with overt indications of American support – but the development of weapon research and production capabilities in Taiwan may reduce the political cost to the United States of maintaining its support for the independence of Taiwan.[8]

Our third point concerns the sale or use of domestically designed or coproduced arms, or the threat of such sale or use in pursuit of economic or diplomatic objectives. The experience of industrialized countries has shown that defense industries impose such high burdens on resources that exports are required to recoup investment costs. Korea seems from the beginning to have had as a deliberate objective the generation of export revenues from its production programs, something which for political reasons has not been as pronounced in Taiwan. The difficulties that new producers face in penetrating markets already dominated by industrial countries provide a strong incentive to seek market outlets among countries with which most Western nations prefer to maintain only distant relations – pariah states, states involved in armed conflicts, or communist-bloc countries.

Defense production capabilities in Taiwan and Korea could influence international politics significantly through arms exports to countries in conflict. As a series of conflicts in the third world has proven, it is often the quantity of even unsophisticated armaments that can accord a decisive advantage to one of the belligerents. Thus, exports by Taiwan or South Korea of even simple munitions such as rocket-lauchers, automatic rifles and flame-throwers could tilt the balance in a conflict. Korea's efforts to export munitions to Iraq in 1981–2 provide a good example of commercial interests dominating political considerations, with potentially grave international consequences.[9] Taiwanese sales to South Africa, though representing only small quantities of support equipment, contravene the internatonal embrago on arms exports to that country. In such transctions, Taiwan's lack of international status works to its advantage: the island has nothing to lose in trading with South Africa.

There are limits, of course, on the extent to which Taiwan and South Korea can contravene US policies without endangering relations with the United States. In addition to assistance and transfers, both countries continue to depend on the United States for what legitimacy they enjoy in the international system. Taiwan has particularly limited latitude; it must balance the potential advantages of tactical maneuvers against the real threat that the United States might make further cuts in exhanges in response to serious violations of US dictates. Korea has more latitude, because the United States cannot lightly dispense with the strategic advantages of its presence in Korea, nor view with equanimity the possibility of instability on the Korean peninsula. Armed and stocked at present levels, both countries can nevertheless make their views felt in bilteral and multilateral diplomatic exchanges. We shall return to this point in the next section.

A fourth way in which the development of military industries has increased the diplomatic security of Taiwan and South Korea is through the growth of relations of trade and assistance that are much deeper than those that in the past accompanied the simple transfer of relatively unsophisticated armaments. For Taiwan, the development of defense-industrial capabilities, achieved largely through coproduction and codevelopment programs with the United States, has deepened, in certain ways, American involvement in the future of Taiwan's defense. This is an interesting paradox in light of the ostensible intent of these programs to strengthen indigenous capabilities and permit the withdrawal of US military forces. However, coproduction programs require long-term commitments and arouse protective sentiments among American industrialists. Furthermore, the United States has a clear stake in avoiding actions that could destabilize the Taiwanese economy, as would have been the case, for instance, if the F-5 aircraft production program at Tai-chung had been terminated. Although assessments differ, it is generally understood that if tighter constraints were placed on Taiwan's access to US technology, it would force that country to divert increasing amounts of national resources into the defense sector. The additional costs of accelerating indigenous research and development and of augmenting defense production plants beyond existing capacities could so severely tax the domestic economy that civilian investments, economic growth and present levels of domestic welfare would be endangered. Mobilizing national resources for defense production in this fashion would require draconian measures that would certainly interrupt any further moves toward political liberalization and would provoke political unrest. If these problems became severe enough, they could affect the region as a whole, tempting the PRC to attempt to improve its own position through subversion or direct military attack.

The balance of forces in US–PRC–Taiwan relations is fragile. At the same time, the existence of strong US–Taiwan bilateral interests, increasingly based on economic factors that are not nearly so mutable or transitory as political or military considerations, provides a margin of safety for the island. It seems to be understood generally in US policy circles that any exploitation of closer US–PRC relations aimed at exerting pressure on the Soviet Union must achieve these ends without provoking an irreversible rupture in US–Taiwan relations or inducing the perception among other Asian states that the United States is willing to sacrifice traditional allegiances in quest of short-term geostrategic advantages.

The same complex interdependency characterizes relations between Korea and the United States. The ostensible US objective of fostering self-reliant forces in Korea (as in other client states) has been contradicted by the simultaneous impulse to export even more advanced equipment requiring larger numbers of American advisors and greater financial support to maintain. Systems presently scheduled for procurement by Korea will require still more human and financial resources. The continuing American presence, and the Korean perception that the Americans are at least partially to blame for its present defense difficulties, make it likely that the United States will be asked to play an even more active role in Korea's defenses, the objective of self-reliance notwithstanding.

Unlike Taiwan, Korea has encountered only sporadic resistance in its efforts to acquire advanced technology to supplement domestically produced arms. The country's effort to acquire the F-16 is a case in point. Korea made repeated attempts to acquire this aircraft in the 1970s, but was turned down or deferred by three successive American administrations. At the time Chun assumed power, however, the United States was willing to approve the sale. Simultaneously, the Chun government sought a license with the United States to coproduce the F-5E. The financial difficulty of supporting both aircraft programs triggered considerable internal opposition among Korean planners. At the same time, Chun could not sacrifice the political benefits of being the first Korean leader to negotiate successfully for the F-16. As a result, he opted for both programs and the United States agreed. The result will be an extraordinary financial burden for Korea and may eventually require either termination of the F-5E coproduction (sacrificing the technological and industrial benefits associated with it) or reduction in the number of F-16s to be purchased. In any case, it is clear that Korea has entered into a force modernization program that may be more important symbolically than it is militarily.

The United States could choose to play a more intrusive role in assisting Korea to allocate defense resources more efficiently, to avoid the repetition of past mistakes, and to maximize the military utility of its production programs. To pursue such a course, however, would require the United States to assume overt responsibility for the direction of Korean modernization programs, thus further exposing the country to blame in the event of mistakes. It is not at all clear that it is in the US interest to increase, rather than decrease, its involvement in Korean defense planning, since the capacity to produce arms domestically and to acquire and adapt needed technologies from many suppliers has reduced its ability to regulate Korea's conduct, conduct which has the potential of implicating the United States and its military forces (if, for example, South Korea embarked upon offensive military action).

From the standpoint of US military objectives for Korea, the present policy guiding technology transfers, which ostensibly aims at bolstering Korea's independent military production effort, does not seem to be based on a clear understanding of the ways the United States would like to see the country develop. Self-reliance is an ambiguous abstraction that clearly does not even approximate a realistic basis for American objectives vis-à-vis Korean military modernization. Contradictions abound. The United States wants Korea to produce munitions, for instance, but will not tolerate commercial competition from Korean exports. The United States wants Korea to be more capable in the production of a range of systems, but disapproves a significant amount of what Seoul's planners have indicated to be ultimate military objectives. Such contradictions need more active discussion if they are to be prevented from becoming a source for serious friction in bilateral relations. The fact that Korea has circumvented US guidelines in the past – both through investment in undesirable programs and through unapproved exports – points to the likelihood of future friction between the two countries as Korea's technological abilities grow.

Military Security

A nation's forces will be effective to the extent that available materiel is appropriate to military strategy. That strategy, in turn, must take accurately into account the political, economic and diplomatic limitations of the nation. The ability to pursue regional military aspirations through independent offensive action is still a fairly distant possibility for both Taiwan and South Korea, given the still limited capabilities of their defense industries, from which neither country has, to date, gained significant quantitative or qualitative advantages over prior levels of imported arms. It is important to keep in mind, however, that defense manufacturing, unlike importing, is a dynamic process that, over time, may yield significant breakthroughs. The simple existence of manufacturing plants and expertise permits countries to exploit new avenues for technological improvements – such as acquisition of dual-use technologies or the clandestine acquisition of required inputs – that could allow the previously dependent state to escape the control of its former patrons.[10] It is entirely possible that Taiwan or South Korea could find reasons to concentrate resources toward the development of types of technologies that would permit them to develop new offensive capabilities. Much has been written about the two countries' nuclear programs, for example, often at the expense of analysis of the conventional technologies that are also potentially significant. So far, for the most part, these latent capabilities to advance both nuclear and conventional programs of which the United States does not approve have been used as political weapons, mechanisms for gaining concessions from the United States. Over time, however, they could be exploited militarily. Additional shrinking in US ties with Taiwan – or even an indication that this was imminent – could provoke Taiwan's leaders to seek radical military solutions in a desperate effort to alter the regional balance. This would be especially likely if the disruption of ties came during a period of economic trouble. Under certain conditions, the leaders of South Korea could also seek new weapons for use against the North.

For the time being, defense production plants in some respects constitute a military liability. Likely targets in the event of conflict, their capture or destruction would be a psychological defeat at the very least. As capabilities become more advanced, moreover, they may invite preemptive attack. This logic has been used to deter Taiwan and South Korea from developing nuclear capabilities, and it can be applied as well to advanced conventional technology such as ballistic missiles.

From a political standpoint, defense-industrial capabilities have added a qualitatively new facet to existing political–military rivalries. Perceptions by opponents of each other's production capabilities, and the consequences these industries are thought to pose for the relative balance of forces, may serve to heighten regional tensions, replicating the dynamics of traditional arms races in which states take steps to improve their own forces, spurred on by exaggerated perceptions of the enemy's strength. Taiwan, for instance, maintains as official dogma a tremendously overdramatized view of the reverse engineering abilities of the communist Chinese. This alleged capability is often used as a rationale for the acquisition of new technologies from the United States. Similarly, Korea, under President Park, set out on a determined course to match and exceed production capabilities attributed to North Korea.

The responsibilities of planning the production of a good part of a nation's military equipment, and coordinating it with necessary imports of advanced technology within the context of an overall military and industrial strategy, might logically be expected to lead that nation's leaders toward an assessment of defense requirements based upon realistic appraisals of the external threat facing the nation. In practice, of course, the perception of threat and the manner in which that perception is applied to the procurement of specific weapons systems is vitiated by a number of powerful factors. These include ideology, economic interest, institutional (for example interservice) rivalry, and the rarely disinterested blandishments of dominant alliance partners.

For years, Taiwan has harbored notions about its defense requirements based on unrealistic perceptions of its national mission. The idea that Taiwan would one day recapture the mainland required certain types of weapon and troop deployments which by any objective measure were not particularly efficient for the defense of the island. In particular, this *idée fixe* gave undue emphasis to the role of ground forces. Under the direct influence of Chiang-Kai Shek and other political and military leaders who espoused this line, Taiwan consistently pressed for an ever larger army. It was not until the 1960s that the Taiwanese air force and navy began to receive modern weapons. The disparity in emphasis among the services, in spite of great strides toward modernization, persists even today.[11]

Overall, the need to produce weapons to substitute for US supplies is having a salutary effect on domestic defense planning in Taiwan to the extent that it has forced scrutiny of defense allocations and required much more careful attention to the feasibility of alternative force postures. Even in the current environment, however, efforts to introduce alternative contingencies into the debate on military strategy meet with resistance. It was not until a few years ago that the discussion of a naval blocakade by the PRC – certainly a realistic and cost-effective way for the PRC to bring pressure to bear on Taiwan – was even permitted. With the tremendous constraints under which Taiwan's defense planners now must operate, it is becoming increasingly difficult to engage in costly 'prestige' projects at the expense of less visible, but more efficient, procurement options. This has expanded the influence of decision-makers possessing the requisite technical and tactical expertise to make judgements among alternative options.[12] It nevertheless has not ended the tension between the pragmatists and defense ideologues; the latter so-called political commissars still enjoy a large measure of influence.

Taiwan's defense capabilities depend critically on the ability to exploit qualitative superiority in weapons, superior troop training, tactics, morale and geographic advantages. Thus decisions about production and procurement of sophisticated weaponry are extremely sensitive. Even one mistake – a misbegotten project that fails – could have severe effects on Taiwan's limited resources and defense options.

The combination of severly limited resources and intense divisions among defense planners creates a highly-charged decision-making environment. The structure of defense management in Taiwan can be affected rather negatively by these forces. Although procurement is officially based on a sophisticated American system in which costs and benefits are carefully weighted, decisions are, in practice, more dependent on political factors than most Taiwanese like to admit. The selection of weapon systems, for instance, can be disrupted totally by a sudden decision to spend resources on a previously unplanned acquisition. The purchase of Dutch submarines in 1981, which by most estimates was a political expendient rather than a rational force-modernization measure, required complete revision of planned expenditures.

Nevertheless, Taiwan's production priorities have generally emphasized the defense forces which the island needs most critically – advanced aircraft and missiles to maintain a margin of superiority in the event of an attack by the People's Republic. This has occurred with the understanding that future purchases in this area may be restricted. The island has deliberately avoided engaging in production projects that might excessively tax its financial and technical and capabilities and drain away resources needed for more immediate defense needs.

Korea's defense planning apparatus is more enigmatic than is Taiwan's because it is rarely discussed openly, and because for years decisions were made deep within the Blue House. Thus, the struggle among the services for allocations, the divisions among different sectors of the economy, and even the internal debate over strategy are hard to decipher.[13] The one constant is the country's antagonistic competition with the North. Korea's major objective has been and will continue to be to gain military superiority over the Democratic People's Republic of Korea, an aspiration that has often exceeded the bounds of what the United States considered to be in Korea's interest or in the interest of regional stability.

It would appear that Korea's investment in defense production has not had the tempering influence over defense planning that is evident in Taiwan. The Korean government has apparently never considered trading off certain imports in favor of indigenous production, or vice versa. In its quest to generate quick export revenues and make quantum leaps in the production of weapons not available through imports, Korea has tended to expand its military sector (like its industrial sector) without weighing too seriously the effect that such an expansion could have on Korean economic performance in overtaxing its human and capital resources. The ability to obtain loans and foreign capital has become a permanent feature of Korean planning and, as such, only externally imposed constraints limit its expansion.

Park's defense industrialization efforts, often more symbolic than instrumental, have not moved Korea toward more realistic assessments of its defense needs. Fractured and subordinated to personal political influence, the planning apparatus in Korea remains weak. In particular, in spite of the considerable expansion of the defense sector during Park's reign, with commensurate growth in the number of advisors and analysts associated with the Ministry of Defense, there seems to have been little integration of military experts into the decision-making apparatus. Defects in Korean-produced equipment undoubtedly helped to reinforce the impression in the military that the Korean leadership was inept and unprofessional, unconcerned about troop safety and morale. Park Chung Hee fell victim to his own ambitious plans.

Taiwan and Korea still depend on the continued commitment of the United States in order to sustain credible security postures.

Nevertheless, changes in the nature of bilateral relations afford increasingly less decisive influence to the United States over the two countries' policies. In the past, these state were virtual hostages in a global security system forged around US containment policy and, given their serious underdevelopment and military insecurity, were unable to act autonomously. Today, the military technological capabilities of both states have become prominent in their diplomatic repertoires, even though true self-reliance remains a chimera.

Notes

Chapter One: Introduction: Defense in the Dependent State

1. If the calculation of the inefficiencies of defense industries were to be the generally accepted basis of national planning, far fewer *industrial* countries would attempt to compete in the production of certain types of advanced military technologies.

2. An interesting analysis of Japan's prewar economy supports the essential hypothesis that a country's development pattern can be determined by its perception of external threat. Norman Buchanan and Howard S. Ellis describe Japan's transformation in the half-century before 1914 from a stagnant and isolated society into a dynamic industrial society as being a result of the fear of its leaders (newly come to power in the Meiji restoration of 1868) that without rapid industrial growth the country would fall under the domination of a foreign power. See Buchanan and Ellis, *Approaches to Economic Development* (New York: Twentieth Century Fund, 1955) p. 181. E. S. Browning has advanced an argument about South Korea's economy that also supports the hypothesis presented here. According to Browning, it was the Nixon Doctrine and the anticipation of troop withdrawals that led South Korea in 1973 to begin a crash program of rapid industrialization in heavy and chemical defense-related industries. See E. S. Browning, 'East Asian Economies', *Foreign Affairs* (Fall 1981) 123–47. A full discussion of the views of Browning and others is presented in Chapter 4 of this text.

3. For an exposition of the unbalanced growth or 'trickle-down' model, see Benjamin Higgins, *Economic Development: Principles, Problems and Policies* (New York: W. W. Norton, 1959) and Gerald M. Meier, *The International Economics of Development* (New York: Harper & Row, 1968) chs 7–9.

4. See, for example, W. W. Rostow, 'The Take-off into Self-sustained Growth', *Economic Journal* (March 1956) 422–63.

5. The principal critics of 'trickle-down' theory include W. H. Singer in, for example, 'The Distribution of Gains Between Investing and Borrowing Countries', *American Economic Review, Papers and Proceedings* (May 1950); Raul Prebisch in the United Nations Report on the Conference on Trade and Development, *Towards a New Trade Policy for Development*

139

(New York, 1964); and Gunnar Myrdal, *Rich Lands and Poor* (New York: Harper & Row, 1957).

6. See, for example, Lewis T. Wells (ed.), *The Product Life Cycle and International Trade* (Boston: Division of Research, Harvard Graduate School of Business Administration, 1972). The product life cycle theory has particular relevance to the study of technology transfer from developed to developing countries, elucidating the respective objectives of supplier and recipient. For the recipient, the fundamental objective is to achieve, over time, the ability to control the technological process in question so as to gain economic autonomy and/or bargaining power. For the supplier, the goal is to protect for as long as possible a competitive advantage over the recipient so as to ensure economic returns, and, in the case of defense technology, political influence as well. As one analyst described the supplier role: 'Technology transfer is a series of deals through which [the supplier] is progressively selling this competitive advantage while the technology itself becomes mature' (Philippe Lasserre and Max Boisot, 'Strategies and Practices of Transfer of Technology from European to ASEAN Enterprises', Institut Européen d'Administration des Affaires (April 1980) unpublished, p. 27). The ultimate measure of control of technology for the recipient is the ability to conduct basic and applied research, as discussed in Chapters 3 and 4 of this text.

7. Stephanie Neuman supports the idea of positive spin-offs to civilian growth from defense programs in her theoretical discussion of arms transfers and economic development in 'Arms Transfers and Economic Development: some Research and Policy Issues', in Stephanie G. Neuman and Robert E. Harkavy, *Arms Transfers in the Modern World* (New York: Praeger, 1979) pp. 246–63. For more empirical observations, see Stephanie G. Neuman, 'Security, Military Expenditures and Socio-economic Development: Reflections on Iran', *Orbis* (Fall 1978) 588–92.

8. For a broader discussion of defense employment issues in advanced countries, see W. Leontieff and M. Hoffenberg, 'The Economic Impact of Disarmament', *Scientific American* (April 1961) 9. On this and the more general factors that distinguish military from civilian industry, see Jacques Gansler, *The Defense Industry* (Cambridge, Mass.: MIT Press, 1980); Seymour Melman, *The Defense Economy: Conversion of Industries and Occupations to Civilian Needs* (New York: Praeger, 1972); A. M. Agapos, *Government–Industry and Defense: Economics and Administration* (University of Alabama Press, 1975); Richard Oliver, 'Employment Effects of Reduced Defense Spending', *Monthly Labor Review* (June 1971); US Bureau of Labor Statistics, *Projections of a Post-Vietnam Economy, 1975* (Washington, DC: US Government Printing Office, 1972).

9. Samuel P. Huntington, *Political Order in Changing Societies* (New Haven: Yale University Press, 1964); Morris Janowitz, *The Military in the Political Development of New Nations* (University of Chicago Press, 1964); and Lucian W. Pye, 'Armies in the Process of Political Modernization', in John J. Johnson (ed.), *The Role of the Military in Underdeveloped Countries* (Princeton University Press, 1962).

10. See, for example, Herbert Wulf, 'Dependent Militarism in the Periphery and Possible Alternative Concepts', in Neuman and Harkavy, *Arms*

Transfers, or Irving L. Horowitz, 'Military Origins of the Cuban Revolution', *Armed Forces and Society*, 4 (1975) 41.

11. For a fuller discussion of this phenomenon from two alternative perspectives, see Michael Moodie, *Sovereignty, Security and Arms, vol. 7, no. 67 of The Washington Papers* (Beverly Hills and London: Sage Publications, 1979), and Wulf, 'Dependent Militarism', pp. 246–63. A major difference between these treatments of the issue of dependence is that Moodie sees it as a short-term, dynamic and operational problem, whereas Wulf presupposes an immutable condition of dependency for developing countries that is simply exacerbated by efforts to produce modern military technology. Wulf derives most of his insights from the body of socioeconomic theory known by the Spanish word of *dependencia*.

12. Geoffrey Kemp identified this phenomenon as one of the so-called back-end problems that affect arms acquisition programs once a transfer decision has taken place: 'Although most recipients of modern weapons wish to maximize their self-sufficiency for reasons of national control, a majority also wants to have the most up-to-date weapons. This can lead to a vicious circle: as a country reaches the point of becoming relatively self-sufficient in operating a given weapon system, a more modern weapon is procured which requires a new series of dependencies upon the supplier.' See Kemp, 'Arms Transfers and the Back-End Problem in Developing Countries', in Neuman and Harkavy, *Arms Transfers*, p. 270. Countries endeavoring to achieve greater independence in force planning may experience a net loss in military capabilities unless they continue to import advanced weapons they cannot yet produce.

13. For case histories of changes in the bargaining power of unequal states as the sophistication of the host country and the sunk costs of foreign investors grow, see Theodore H. Moran, *Multinational Corporations and the Politics of Dependence: Copper in Chile* (Princeton University Press, 1974); Franklin Tugwell, *The Politics of Oil in Venezuela* (Stanford University Press, 1975); Constantine V. Vaitsos, *Intercountry Income Distribution and Transnational Enterprises* (Oxford: Clarendon Press, 1974).

14. As is discussed in more detail in the next chapter, this could be accomplished in one of three ways: (a) by diverting technical inputs or expertise provided for an approved program to the new design and production, (b) by adapting 'dual-use' technologies acquired through commercial channels, or (c) by acquiring necessary inputs from peripheral suppliers clandestinely. See, for example, Science Applications, Inc., 'The Use of Dual-Use Technologies for LDC Arms Production Programs', Report prepared for the US Arms Control and Disarmament Agency, contract no. AC8WC122 (Washington, DC, 1979).

15. Pakistan has translated the threat of nuclear development into advantageous transactions for conventional armaments. The literature on nuclear proliferation provides other interesting examples. See, for example, Joseph Nye, *Foreign Policy* (Spring 1980) 603–23, and Stockholm International Peace Research Institute, *Nuclear Energy and Nuclear Weapon Proliferation* (London: Taylor & Francis, 1979) chs 9, 10.

16. The role that Western advisors are thought to have played in the growth

of dissent in Iran is quite different. Clearly, a pronounced presence of Western personnel, including military personnel, can serve as a focus of discontent for groups that are, for one reason or another, resisting the influence of Western practices, policies or attitudes. In Iran, for example, the existence of large American enclaves in the country served as a partial catalyst for the discontent expressed by traditional groups with the modernizing policies of the Pahlavi régime. At the same time, their presence may have deterred a social revolution from occurring earlier. In South Korea, the presence of Japanese officials, businessmen and visitors is a profound irritant to segments of the population. For a recent assessment of this sentiment, see Tracy Dahlby, 'Ancient Enmities Cast Shadow on South Korea's Ties with Japan', *Washington Post*, 29 January 1982.

17. Guy J. Pauker, *et al.*, *In Search of Self-reliance: US Security Assistance to the Third World under the Nixon Doctrine*, Report prepared for the Advanced Research Projects Agency, grant no. R-1092-ARPA (Santa Monica, Ca.: Rand Corporation, June 1973) p. v. A more recent analysis of this issue can be found in the seminal work of Steven Canby and Edward Luttwak, 'The Control of Arms Transfers and Perceived Security Needs', Report prepared for the US Arms Control and Disarmament Agency, ACDA contract no. AC9WC112 (Washington, DC: C&L Associates, June 1979). For a Marxist model of the radical reconfiguration of third world armed forces, see Herbert Wulf, 'Dependent Militarism', pp. 246–61. Wulf suggests six guiding principles for third world militaries: (a) never counter aggression by using modern, sophisticated major weapon systems; (b) guarantee the effectiveness of a labor-intensive defense system through a high degree of participation by the population; (c) relinquish all modern weapon systems to remove complex logistical requirements and thereby increase independence; (d) decentralize forces (in a people's militia system); (e) foster lightly-armed infantry units, possibly with a small navy for coastal protection; and (f) create a defense system – building on the previous five principles – that is entirely defensive in nature and totally unsuitable for offensive operations.

18. The analysts argued that 'instead of the complex aircraft, artillery, and tracked vehicles provided by the United States in the past, Third World countries need cheaper aeronautical systems, ground-force weapons and mobility keyed to their own needs. A broadly based light infantry force could be substituted for the heavy infantry systems initially designed for the American Expeditionary Force in Europe. Air and naval forces should become secondary concerns' (Pauker *et al.*, *In Search of Self-reliance*, p. vi).

19. There is some evidence to suggest that both states have had development programs for long-range SSMs for some time. This is discussed in Chapter 3 of this text. The potential for adapting to nuclear use delivery systems developed under these programs is also discussed in that chapter.

20. Emile Benoit, with Max E. Millikan and Everett E. Hagen, *Effect of Defense on Developing Economies*, 2 vols, Arms Control and Disarmament Agency report no. E-136 (Cambridge, Mass.: Center for International Studies, Massachusetts Institute of Technology, June 1971).

21. On the basis of a series of case studies – the most important of which was a case study of India – Benoit identified a series of positive and negative effects of military spending for economic growth that would account for the apparent correlation of growth in the gross domestic product with the defense budget. Major variables thought to account for positive effects included an increase in the inflow of foreign aid (correlated with a higher defense burden), the use of the armed forces for public works, and the provision of substitutes for civilian goods and services by the military (sparing the civilian economy the need to provide them). See Benoit, *Effect of Defense*, pp. 278–80.

22. Benoit's findings have been attacked widely by a variety of scholars. Many of these critiques are mere polemic, but others are efforts to point out methodological flaws. Major problems in the study arise from Benoit's effort to conduct a macrostatistical inquiry on a subject that seems to elude cross-country multiple regression analysis. For representative discussion, see Michael Brzoska and Herbert Wulf, 'Rejoinder to Benoit's "Growth and Defense on Developing Countries" – Misleading Results and Questionable Methods', mimeograph (Study Group on Armament and Underdevelopment, University of Hamburg, 1979); Ricardo Faini, Patricia Arnez and Lance Taylor, 'Defense Spending, Economic Structure and Growth: Evidence Among Countries and Over Time', unpublished (Cambridge, Mass.: Massachusetts Institute of Technology, October 1980); and Nicole Ball, 'Defense and Development: a Critique of the Benoit Study', unpublished (Swedish Institute of International Affairs, May 1982).

23. International Institute for Strategic Studies, *The Military Balance 1981–1982* (London, 1981) p. 103.

24. Arthur Alexander, 'Economic Motivations in International Arms Production and Transfers: an Integrated Framework to Analyze Policy Alternatives' (Harvard University, 16 May 1980).

25. Highly advanced ballistic missile technology, certain state-of-the-art technology such as fuel–air explosives, and so-called weapons of ill-repute such as napalm and other incendiaries, are subject to almost universal export prohibition in the United States.

26. Sydney Verba, 'Some Dilemmas in Comparative Research', *World Politics*, 20 (October 1976) 113.

27. Alexander L. George, 'Case Studies and Theory Development: the Method of Structured, Focused Comparison', in Paul Gordon Lauren (ed.), *Diplomatic History: New Approaches* (New York: Free Press, 1979) pp. 46–7.

Chapter Two: The Role of US Policy in Promoting the Defense Industries of South Korea and Taiwan

1. An often-cited example of the unintended effects of a policy pronouncement is Secretary of State Acheson's 'failure', in a speech in January 1950, to mention South Korea (and Taiwan) as areas falling within the defense

perimeter of the United States. For the representative discussions of this period, see A. Doak Barnett, *China and the Major Powers in East Asia* (Washington, DC: Brookings Institution, 1977), or Stephen P. Gibert, *Northeast Asia in US Foreign Policy*, Washington Papers, vol. VII (Beverly Hills and London: Sage Publications, Georgetown Center for Strategic and International Studies, 1979) pp. 37–40.

2. For a cogent discussion of this period in US diplomatic history, see Grant E. Meade, *American Military Government in Korea* (New York: King's Crown Press, Columbia University, 1951) and Frank Baldwin (ed.), *Without Parallel: the American–Korean Relationship since 1945* (New York: Random House, 1974). Stephen Gibert has argued that: '[C] ontainment was not viewed in these early years either in an Asian context or as a perfectly natural corollary of balance of power politics in a bipolar world . . . it was not fully recognized that Soviet–American interactions on the Korean Peninsula were essentially of the same character as in Europe and the Middle East.

    'In Asia as well as Europe, the imperative to prevent a hegemonial position from being established by a hostile power was (and is) a prescribed 'rule' of balance of power politics, made more cogent at the time by the destruction of all rivals to Soviet and American preeminence. Failing to see this, Washington also failed to see the division of Korea as but yet another facet – albeit one geographically far distant from American past involvement and psychologically light years away from the previous isolationist age of innocence – of the developing cold war. Not recognized as such, Korea was an opening phase of a worldwide power struggle with the Soviet Union' (Gibert, op. cit., p. 38).

3. In December 1949, the Department of State issued a classified international memorandum that forecasted the imminent fall of Taiwan to communist forces. See Ralph N. Clough, *Island China* (Cambridge, Mass.: Harvard University Press, 1978) p. 7. By May 1949, Chinese communist forces reportedly had 300,000 assault troops positioned opposite Taiwan for sea-lift in motorized junks and transports. See Edwin K. Snyder, *et al.*, *The Taiwan Relations Act and the Defense of the Republic of China* (Berkeley: Institute of International Studies, University of California, 1980) p. 1; and Barnett, op. cit., p. 176.

4. On 27 June 1950, President Truman declared that 'in these circumstances the occupation of Formosa by Communist forces would be a direct threat to the security of the Pacific area. The determination of the future status of Formosa must await the restoration of security in the Pacific, a peace settlement with Japan, or the consideration by the United Nations' (*State Department Bulletin* (3 July 1950) 5–6, quoted in Barnett, op. cit., p. 376).

5. The Korean War also entrenched US hostility to China after Chinese troops moved in to bolster North Korean forces following the crossing of the 38th parallel by UN forces, in effect rendering the Korean conflict a China–US conflict from late 1950 to mid-1953. Previous to Chinese involvement, however, the United States had already moved significantly toward a comprehensive policy of anti-Communism, with China as major nemesis. For a detailed account of this period from the standpoint of US domestic politics, see John King Fairbank, *The United States and China*

(Cambridge, Mass.: Harvard University Press, 3rd ed., 1972) pp. 315–19. For a more general discussion of this period of US foreign policy interests in Asia, see Fred Greene, *US Policy and the Security of Asia* (New York: McGraw-Hill, 1968).

During the Korean War, US military spending had increased from $13 billion to $50 billion a year. The size of the armed forces in 1950 (1,461,000) doubled before 1952, as did the number of Air Force wings, which jumped from 48 in 1950 to 95 in 1952. In addition, the United States had established a chain of island bases in Japan, Okinawa, Taiwan, the Philippines, Australia and New Zealand, comprising the major basis of the American forward defense line in the Far East. See Samuel P. Huntington, *The Common Defense* (New York: Columbia University Press, 1961) pp. 52–64.

6. For a more detailed discussion of the nature of political relations with the US, for Taiwan, see King-yuh Chang, 'Partnership in Transition: a Review of Taipei–Washington Relations', *Asian Survey*, XXI, 6 (June 1981) 603–20; Hungdah Chiu (ed.), *China and the Taiwan Issue* (New York: Praeger, 1979); for Korea, see, for example, Nena Vreeland *et al.*, *Area Handbook for South Korea* (Washington, DC: US Government Printing Office, 1975) ch. 9.

7. As Edward Mason observed about the American posture vis-à-vis Korea's economy during this period: 'Washington could not make up its mind whether its primary objective was to keep the Korean economy afloat for security reasons or whether economic development was also an important desideratum.' For full discussion of this question, see Edward S. Mason *et al.*, *The Economic and Social Modernization of the Republic of Korea* (Cambridge, Mass. and London: Harvard University Press, 1980); for Taiwan, see Neil H. Jacoby, *US Aid to Taiwan* (New York: Praeger, 1966).

8. The United States dominated investment in Korea and Taiwan throughout the period between 1953 and the early 1970s. In Taiwan, for instance, American investment between 1952 and 1974 amounted to $1,924 million in capital investment, approximately 44 per cent of total foreign capital in that country. South Korea's exports of light manufactures went predominantly to the United States, but exports of primary products were divided more evenly between the United States and Japan. For detailed discussion, see Ramon H. Meyers, 'The Economic Development of Taiwan', in *China and the Question of Taiwan: Documents and Analysis* (New York: Praeger, 1973), and Vreeland, op. cit., p. 347.

9. Starting as early as 1961, the Park régime, relying on a cadre of military élite, launched a series of five-year plans which had as a major emphasis investment in areas of particular utility to military preparedness – petroleum, chemicals, railroads and highways. In the decade after the 1961 coup, former military officers entered the private and public sectors in managerial positions, bringing with them expertise and technical orientation largely gained from their military training under the aegis of the United States. Although this development strategy was much accelerated after 1970 and the Nixon Doctrine, there was, even in the earlier years of American influence, a recognition of the importance of developing an economy able to sustain a pronounced military effort. Taiwan was

rather slower in moving toward heavy industry, relying more on technology-intensive production aimed at maximizing foreign exchange receipts, but this too had as an implicit aim the support of a more self-reliant defense sector. Partly in response to the perception of weaknesses in the US security commitment arising from the American preoccupation with Vietnam, Taiwan as early as the mid-1960s established the Chungshan Institute of Science and Technology to initiate projects for US weapons. See Chapter 3 of this study for more detailed discussion.

10. Throughout the 1950s, a series of incidents occurred involving China and Taiwan and, on occasion, US armed forces, culminating in the Quemoy crisis of 1958, in which US policy-makers went so far as to threaten the use of nuclear weapons. See, for instance, M. H. Halperin, *The 1958 Taiwan Straits Crisis: a Documented History*, RM 2900 ISA (Santa Monica, Calif. Rand Corporation, December 1966).

11. See, for instance, Se-Jin Kim, 'South Korea's Involvement in Vietnam and its Political and Economic Impact', *Asian Survey*, x, 6 (June 1970) 319–33. The American experience in South-east Asia had a profound impact on the security perceptions of Taiwan and South Korea. This is discussed in Chapter 3 of this text. It should be noted in particular that even prior to the withdrawal of American troops from Vietnam, both states experienced relative unease arising from the redeployment of a large part of the Seventh Fleet to South-east Asia. In Taiwan, the US Navy's Taiwan Patrol Force was reduced to a destroyer escort by 1969, certainly serving to underscore Taiwan's desire to replace the diminishing US force with independent capabilities. The overt articulation of this objective did not occur, however, until the beginning of US rapprochement with the People's Republic in 1972. For further discussion of Taiwanese naval development, see William J. Durch, 'The Navy of the Republic of China', in Barry M. Blechman and Robert P. Berman (eds), *Guide to Far Eastern Navies* (Annapolis: Naval Institute Press, 1978) ch. 2.

12. Although there was a gradual decline in US foreign aid (including military assistance grants, overall economic aid, and direct gifts or arms) during this period, the big change in policy did not come until after 1969. Asia, moreover, did not lose concessionary benefits from the US as quickly as did other regions of the world (such as NATO). However, in the period between 1955 and 1965, annual global sales of arms (as opposed to subsidized transfers) rose from about $80 million to more than $1.75 billion, an increase over 2000 per cent. See US Department of Defense Security Assistance Agency, *1979 Fiscal Series 2* (Washington, DC: US Government Printing Office, 1979), and David J. Louscher, 'The Rise of Military Sales as a US Foreign Assistance Instrument', *Orbis* (Winter 1977) 933–64.

13. For a complete discussion of the evolution of forces in both South Korea and Taiwan, see Chapter 3 of this text.

14. See the *Report to the Committee on International Relations, US House of Representatives, by the Comptroller General of the United States: Coproduction Programs and Licensing Arrangements in Foreign Countries* (Washington, DC: US General Accounting Office, 1975).

15. For representative analyses of the Nixon Doctrine and its effects on Asian

security, see the following: Stephen P. Gibert, op. cit., pp. 41–4; Guy Pauker *et al.*, *In Search of Self-reliance: US Security Assistance to the Third World under the Nixon Doctrine*, grant no. R-1092-ARPA (Santa Monica, Ca.: Rand Corporation, June 1973); Yuan-li We, *US Policy and Strategic Interests in the Western Pacific* (New York: Crane, Russak, 1975); and Edwin K. Snyder *et al.*, *The Taiwan Relations Act and the Defense of the Republic of China* (Berkeley: Institute of International Studies, University of California, 1980) ch. 5.

16. See *National Security Strategy of Realistic Deterrence*, Secretary of Defense Melvin Laird's Annual Defense Department Report, FY1973 (February 1972) pp. 23–51.

17. Over 700,000 US troops were withdrawn from Asia between 1969 and 1975. Pursuant to the Shanghai communiqué in 1972, the United States reduced the number of American servicemen on Taiwan from 10,000 to 2000 by 1976. The Seventh Division (one of two divisions deployed in Korea since 1950) was withdrawn in 1971–2. For more detail on Taiwan, see US Congress, Senate, Committee on Foreign Relations, *Taiwan: Hearings on US Policy in East Asia*, 96th Congress, 2nd sess. (Washington, DC: US Government Printing Office, 1979); for South Korea, see Larry A. Niksch, 'US Troop Withdrawal from South Korea: Past Shortcomings and Future Prospects', *Asian Survey*, XXI, 3 (March 1981) 326–41.

18. Stephen Gibert cogently describes the fundamental inconsistency of the Nixon Doctrine in attempting simultaneously to lower the American commitment of troops and decrease concessionary aid: 'Central and critical . . . to this new direction in American policy was the provision of large scale military aid and training to enable local forces to take over the security role being vacated by the United States. This did not occur, however, on the required scale because the Nixon Administration failed to recognize an important, if obvious, fact of political life: as Asianization . . . proceeded, military aid amounts were bound to decline. Intensity of commitment, presence of troops, and level of aid are intimately related . . . a decline in troop presence inevitably meant a lessening of American commitment and that, in turn, meant reduced, not increased military aid' (Gibert, op. cit., p. 43).

19. As a 1973 Rand study put it: 'The [1972 Defense] Report gives the impression that it was thought necessary to avoid a drastic break with that past security policy based on the idea that the United States had to prepare for an ultimate confrontation with the Communist world. Although other interpretations are possible, the obvious one is that the United States wants to retain its role as leader of a major coalition of nations but, at the same time, to reduce its paternalistic role. If that interpretation is correct, the change in foreign and security policies does not point toward pragmatic cooperation with any friendly country eager to maintain its independence and freedom of action, but toward the more limited purpose of redistributing burden-sharing among partners with common goals' (Guy Pauker *et al.*, op. cit., p. 5). Stephen P. Gibert described the strategic re-assessment as a replacement of the 'two-and-a-half war' doctrine with the 'one-and-a-half war' concept: 'Given the Sino–Soviet rift, US forces could be reduced to those sufficient to combat the

Soviet Union in Europe and People's Republic in Asia but need not be prepared to deal with a combined Sino–Soviet assault. In addition, US forces should be able to fight a "half-war" (a local limited conflict) elsewhere simultaneously' (Stephen P. Gibert, op. cit., p. 42).

20. Throughout the early and mid-1970s, a series of Congressional visits to Asia served to keep up active debate over US policy to the region. A representative report of a delegation can be found in US Congress, House, Committee on International Relations, 95th Congress, 2nd Session, *Asia in a New Era* (Washington, DC: US Government Printing Office, 1975).

21. See William E. Griffith, *Peking, Moscow and Beyond* (Beverly Hills: Sage Publications, 1973) pp. 1–5, and Ralph N. Clough, *East Asia and US Security* (Washington, DC: Brookings Institution, 1975) chs 3, 7.

22. As Edwin K. Snyder describes it: 'The American move toward Peking in the pursuit of a balance-of-power policy suggested a dilution of US military guarantees to its allies of the "bipolar" period throughout the Pacific basin, which led them to attempt to accommodate former adversaries, to seek their protection, to at least strike a pass of neutrality' (Snyder, op. cit., p. 9, summarizing arguments presented by Earl Ravenal in 'Consequences of the End Game in Vietnam', *Foreign Affairs*, 53 (July 1975)).

23. See Chapter 3 in this study for further discussion of these measures.

24. This is a hotly contested issue. Some would argue that President Park simply seized upon the opportunity presented by changes in the external security environment to justify draconian internal measures to bolster his political power at a time when support was waning. On the other hand, threats to internal stability arising from rapid changes in perceptions of the strength of US commitments cannot be discounted. As Robert Scalapino has argued: 'At an earlier period, Park had been persuaded to experiment with parliamentarianism, at a time when the United States was strong and credible in his eyes. With the decline of American influence in Asia (and the opening of negotiations with North Korea), Park moved in an authoritative direction in 1972 in an effort to cement national unity' (Robert A. Scalapino, 'Asia at the end of the 1970s', *Foreign Affairs, America and the World, 1979*, 58, 3 (1980)). See Chapter 5 in this study for further discussion.

25. E. S. Browning, 'East Asian Economies', *Foreign Affairs*, 60, 1 (Fall 1981) 128.

26. The only third world coproduction contract traceable to the early 1970s other than in Taiwan and South Korea was an agreement with Iran to retrofit the M-47. See US General Accounting Office, *Report to the Committee on International Relations*, op. cit., p. 20.

27. Robert G. Sutter and William de B. Mills, 'Fighter Aircraft Sales in Taiwan: US Policy', Congressional Research Service, issue brief no. IB81157, 28 October 1981 (updated 20 January 1982).

28. Snyder describes this increase in sales as follows: 'The arms transfer policy identified with Henry Kissinger was predicated on the conviction that global and regional stability was governed by the balance of power and that the transfer of arms, if controlled by the United States, would enable the United States to preserve that balance. It was assumed that the

provision of arms to other nations would contribute to the maintenance of peace and the furtherance of American interests' (Edwin K. Snyder *et al.*, op. cit., p. 64). A full description of the decision-making apparatus for arms transfers and the Congressional statutes which serve as guidelines can be found in the following: US General Accounting Office, Report to the Congress, *US Munitions Export Controls Need Improvement*, ID-78-62 (25 April 1979); David Louscher, 'The Emerging Arms Transfer Decision Process', testimony before the Subcommittee on International Security and Scientific Affairs, Committee on Foreign Affairs, House of Representatives (23 February 1979); and US Arms Control and Disarmament Agency, Report to the Congress, *The International Transfer of Conventional Arms* (Washington, DC: US Government Printing Office, 12 April 1974). These figures are from *Senate Foreign Relations Committee Report on the International Security Act of 1978*, cited in Snyder *et al.*, op. cit., p. 63.

29. See US ACDA Report to the Congress, op. cit., p. 31.
30. *A Staff Report to the Subcommittee on a Foreign Assistance, Committee on Foreign Relations, US Senate, US Military Sales in Iran* (Washington, DC: US Government Printing Office, July 1976).
31. The International Security Assistance and Arms Control Act of 1976 (Public Law 94-329). See the *Report to the Congress on Arms Transfer Policy* (Washington, DC: US Government Printing Office, 1977).
32. See 'Text of the Statement of Conventional Arms Transfer Policy' (issued by the President in May 1977) p. 1.
33. Report to the Congress on Arms Transfer Policy, op. cit.: 1–2.
34. These two decisions were inherited by the Carter administration from its predecessor, however, adding to the pressure to approve the sales in spite of the restraint guidelines. Regardless, these early initiatives served to highlight the fundamental incompatibility between general guidelines for arms sales and the highly volatile and idiosyncratic environment in which such decisions are necessarily taken, where politics usually prevail.
35. Larry A. Niksch, among others, has identified three central indices of the Carter administration's views on Asia which dominated policy formulation in the first year. These include, firstly, the notion that the withdrawal of American ground forces from Asia could preclude American involvement in a land war, which was a primary objective. Presidential Review Memorandum 13 emphasized that the removal of the Second Division from Korea, in particular, would remove the so-called trip-wire which would prompt immediate American involvement in a ground conflict in the event of North Korean aggression. PRM-10 had stated this even more categorically in saying that the removal of US forces in Asia would permit the United States to have 'flexibility to determine at the time whether it should or should not get involved in a local war'. Secondly, there seemed to be widespread support in the administration for a European emphasis in overall strategy, reflected in the so-called 'swing-strategy', which dictated that in the event of US–Soviet conflict, the US would redeploy its Pacific naval forces to Europe. Niksch argues that this 'Europe first' posture led some in the administration to propose ever greater cuts in the US Asian posture, including the removal of one of the aircraft carriers

deployed with the Seventh Fleet, the possible reduction in the use of bases in the Philippines, and the possible withdrawal of the Third Marine Division in Okinawa. Thirdly, it is documented that Carter, on several occasions, revealed a personal preference for de-emphasizing Asia in US foreign policy, reportedly telling Secretary of Energy James Schlesinger that he was determined to avoid the mistakes of his predecessors (including Truman, Johnson and Nixon), all of whom had the effectiveness of their administrations undermined because of preoccupations with Asian wars. See Larry A. Niksch, 'US Troop Withdrawal from South Korea: Past Shortcomings and Future Prospects', *Asian Survey*, xxi, 3 (March 1981), and Donald S. Zagoria, 'Why We Can't Leave Korea', *New York Times Magazine* (2 October 1977) p. 86.

36. See US Congress, House, Committee on International Relations, *Security Issues in the Far East, Report of Fact-finding Mission to South Korea and Japan*, 95th Congress, 1st Session (Washington, DC: US Government Printing Office, 1977).

37. Congressional reaction to the plan was immediate and extensive. It culminated in part in the drafting of legislation that would require consultations with the Congress prior to any further changes in policy toward Asia. It should be noted, however, that segments of Congress in previous years had indicated support for removal of ground forces from South Korea, including the House Defense Appropriations Subcommittee, whose report in 1974 recommended an incremental withdrawal of the Second Infantry Division to begin in Fiscal Year 1976. See Niksch, op. cit., p. 238.

38. One analyst in Korea, Dr Youn-soo Kim, went so far as to advocate that Korea might improve its international standing and its internal power by joining the non-aligned movement, given the high cost of the special relationship with the United States. As he put it: 'Korea must correctly understand from the failure in the Vietnamese War that the best national defense is the prevention of war and the one and only way of national defense is to build its own national power . . . The ROK must establish a national defense system on its own and without support by foreign troops and military bases in Korea' (Youn-soo Kim, 'The ROK, the DPRK and Yugoslavia: 1950–1978', *Korea and World Affairs*, 2, 2 (Summer 1978) 218–46).

39. Report of the Department of State on Security Assistance to Korea, 1978, reprinted in *Department of State Bulletin* (April 1980) 24–5.

40. Ibid., p. 26.

41. President Carter apparently had hoped that China could play a key role in defusing tension in the Koreas so as to permit the continuation of the troop withdrawal plan. In démarches to Deng Xiao Ping in 1979, Carter stressed Korea as a major point of cooperation, hoping that the Chinese would move to recognize South Korea and on that basis begin to build a new framework for restraint of North Korea. Deng flatly refused to consider these proposals. See Larry A. Niksch, op. cit., pp. 333–4.

42. See for instance, Gottfried-Karl Kindermann, 'Washington between Beijing and Taipei: the Restructured Triangle 1978–80', *Asian Survey*, xx, 5 (May 1980) 457–77; Hungdah Chiu, 'The Future of US–Taiwan

Relations', *Asian Affairs* (Winter 1981) 21–7; Chi-wu Wang, 'Military Preparedness and Security Needs: Perceptions from the Republic of China on Taiwan', *Asian Survey*, XXI, 6 (June 1981); and Robert L. Downen, *Of Grave Concern: US–Taiwan Relations on the Threshold of the 1980s* (Washington, DC: Center for Strategic and International Studies, 1981) pp. 1–21.

43. On the subject of arms sales, the US and the PRC simply 'agreed to disagree'. The US maintained that it would sell 'selected defense arms' to Taiwan, while the PRC insisted they would not agree to this. See Harry Harding, Jr, *China and the US: Normalization and Beyond* (New York: China Council, 1979) p. 10.

44. *China Post* (Taipei), 29 December 1978, p. 1.

45. For full discussion of this, see US Congress, Senate, Committee on Foreign Relations, Report 96097, *Taiwan Enabling Act* (Washington, DC: US Government Printing Office, 1979); see also US Congress, Senate, Committee on Foreign Relations, *Taiwan*, 96th Congress, 1st Session (Washington, DC: US Government Printing Office, 1979); Robert Downen, op. cit.; and Edwin K. Snyder *et al.*, op. cit., ch. 2.

46. Section 3 of the Taiwan Relations Act. See Snyder *et al.*, op. cit., app. C.

47. Chi-wu Wang, op. cit., p. 659.

48. See Downen, op. cit., p. 19.

49. Issue Brief no. 1B811 57, op. cit., p. 2.

50. Ibid.

51. The PRC implicated the United States in the sale of two Dutch submarines to Taiwan in December 1980, which resulted in US – PRC friction (in spite of a categorical denial of any responsibility by the United States) and led to the downgrading of diplomatic relations with the Netherlands.

52. Taiwan has received equipment from Israel, including Gabriel and Shafrir SSMs, and has approached South Africa to discuss possible weapon sales. Purchases from Israel seem to be limited, however, in the interest of better relations with Arab countries, which have grown in importance in recent years. Consideration of the Kfir fighter aircraft was apparently abandoned for this reason. See Andrew Pierre, *The Global Politics of Arms Sales* (Princeton University Press, 1982) p. 216.

53. For full discussion of this, see, for instance, US Congress, Senate, Committee on Foreign Relations, *Taiwan: One Year after United States–China Normalization*, 96th Congress, 2nd Session (Washington, DC: US Government Printing Office, June 1980).

54. Interviews with officials at the Department of State, US Arms Control Agency and the Department of Defense did not shed much light on this question. It appears that minor military items are being approved but all high visibility items are, for the moment, 'under review'. It is probably safe to assume that any major sales would have been reported in the media.

55. Department of State, Munitions Control Newsletter, no. 90 (August 1981).

56. Interviews at AIT, Taipei (1982).

57. Ibid.

58. See Julian Weiss, 'Taiwan's Economy in the 1980s; Prospects and Pitfalls

Ahead', in US Congress, Senate, Committee on Foreign Relations, *Taiwan: One Year after United States-China Normalization* (Washington, DC: US Government Printing Office, June 1980) p. 88.

59. For full discussion of the internal political changes that preceded and accompanied derecognition, see, for instance, J. Bruce Jacobs, 'Taiwan 1978: Economic Successes, International Uncertainties', *Asian Survey*, XIX, 1 (January 1979) 21–9; James Hsiung, 'The Security of Taiwan and US Policy', in *Taiwan: One Year After*, op. cit., pp. 121–4; and Jurgen Domes, 'Political Differentiation in Taiwan: Group Formation within the Ruling Party and the Opposition Circles 1979–1980', *Asian Survey*, XXI, 10 (October 1981) 1012–15. The elections scheduled for 23 December were for 53 delegates to the National Assembly and 52 seats in the legislative Yuan. The role of opposition candidates in the election campaign was unprecedented, and included the open announcement of a political platform for non-KMT candidates. This was a degree of liberalization for Taiwan which had never before been in evidence, prompting one observer to say that Taiwan 'had reached the threshold of a multi-party system' (Domes, op. cit., p. 1012).

60. Known as the Kao-Hsiung riots, the disturbances were led by opposition forces centered on a journal known as 'The Magazine of Taiwan's Democratic movement', which had emerged in August 1979 as the supposed mouthpiece of all of Taiwan's opposition factions. For different accounts of the riot, see Phil Kurata, 'Violence in the Name of Reason', *Far Eastern Economic Review*, 28 December 1979, p. 27; James Hsiung, 'The Security of Taiwan and US Policy', op. cit., p. 123; and Domes, op. cit., p. 1013.

61. Some Koreans asserted that the American CIA was responsible for Park's death; see Franklin B. Weinstein and Fuji Kamiya, *The Security of Korea: US and Japanese Perspectives on the 1980s* (Boulder, Col.: Westview Press, 1980) p. 3.

62. Statement by Selig S. Harrison, in Weinstein and Kamiya, op. cit., p. 72.

63. Korea has been attempting for some time to use its relationship with Japan to secure a six billion dollar aid package. This was presented to the Japanese as a form of moral obligation in repayment of the benefits Japan allegedly receives from stability in Korea. The Japanese offered $800 million in aid. See, for instance, Henry Scott Stokes, 'Negotiations on Japanese Aid to Seoul Seem Near Impasse', *New York Times*, 19 April 1982), p. 6.

Chapter Three: The Defense Industries of Taiwan and South Korea in Detail

1. Studies of defense production in developing countries typically lament the absence of reliable data. Information is in fact available, albeit in no organized or comprehensive form. Among the sources used in this study are the 'trade journals', such as *Ground Defense/Armies and Weapons* and *Aviation Week and Space Technology*, reports issued by private consulting

concerns such as the *Defense Market Survey*, congressional hearings and reports, unclassified government documents and studies (US and foreign), private interviews with members of the defense and industrial communities in Taiwan, South Korea and the United States, and, in a few notable cases, unpublished papers and dissertations. This 'all source' research effort clearly could present data comparability problems were this a cross-country comparison of a large number of cases. As a basis for two case studies, however, this eclecticism works. Wherever possible, at least two sources have been sought to verify the more sensitive material. Most studies of the defense production potential of developing countries rely on data published by the Stockholm International Peace Research Institute's *World Armaments and Disarmament* (London: Taylor and Francis, annually), which includes a register 'of indigenous and licensed production' in third world countries. Other general sources of information include Peter Lock and Herbert Wulf, *Register of Arms Production in Developing Countries* (Hamburg: Study Group on Armaments and Underdevelopment, 1977), *Strategic Survey 1976* (London: International Institute for Strategic Studies, 1977), and the US Arms Control and Disarmament Agency's *World Military Expenditures and Arms Transfers, 1969–79* (Washington DC: US Government Printing Office, 1982). These data are, at most, inventories of production programs in existence (for the most part, selective inventories), and provide very little detail beyond the dates of the inception of the program and estimates of the number of systems involved. ACDA provides information on the value of third world exports only. None of these data sets provides a definitive basis for understanding the scope and nature of defense production programs.

2. A discussion of the stages common to most developing country production programs can be found in Science Applications, Inc., 'The Role of Coproduction and Dual-use Technology in the Development of LDC Arms Industries', unpublished draft (September 1979) pp. 20–4; Michael Moodie, *Sovereignty, Security and Arms* (Beverly Hills and London: Sage Publications, 1979) p. 26; and Stephen E. Miller, 'Arms and the Third World: the Indigenous Weapons Production Phenomenon', unpublished paper prepared for the Programme for Strategic and International Affairs, Graduate Institute of International Studies, University of Geneva (June 1980) p. 6.

3. Recipients can, in some instances, by-pass some restrictions by procuring so-called dual-use technology which can be purchased through commercial channels but which has potential military application. Examples of these technologies include older guidance and propulsion systems for use in ballistic missile development. See Science Applications, Inc., *Technology List for Observing Possible Indigenous Development/Production of a Surface-to-Surface Missile System by a Less Developed Country (LDC)*, report prepared for the US Arms Control and Disarmament Agency, no. AC8WC122 (Arlington, Virginia, April 1979), and Mark Balaschak, *et al.*, 'Assessing the Comparability of Dual-Use Technologies for Ballistic Missile Development', report prepared for the US Arms Control and Disarmament Agency, contract no. ACOWC113 (June 1981). The significance of this procurement channel – either from the standpoint of enhancing defense

production programs or in the extent to which this has been utilized in the past by developing countries – has not been established with an adequate empirical support.

4. For a full discussion of licensing agreements and US policy in this area, see Chapter 2 and US General Accounting Office, *Coproduction and Licensing Arrangements in Foreign Countries*, 94th Congress, 2nd Session (2 December 1975) Part II. Perhaps the most important restriction placed on developing country producers is that relating to exports. For the most part, coproduction agreements have a blanket prohibition on re-export of equipment containing US components unless explicit US permission is granted.

5. Chi-wu Wang, 'Military Preparedness and Security Needs: Perceptions from the Republic of China on Taiwan', *Asian Survey*, XXI, 6 (June 1981) 661.

6. William J. Durch, 'The Navy of the Republic of China', in Barry M. Blechman and Robert P. Berman (eds), *Guide to Far Eastern Navies* (Annapolis: Naval Institute Press, 1978) p. 227.

7. Requests from the Republic of China for submarines, for instance, were deferred or ignored as early as 1960, given US reluctance to provide Taiwan with weapons which could be viewed as provocative by the PRC. Subsequent requests in 1968 and 1969 also went unfulfilled, and it was not until 1972 that Taiwan was able to acquire submarines, in this case former US Navy *Tench*-class patrol submarines commissioned in 1945 and 1946. As Durch described US policy at the time: 'The submarines were transferred to enhance Chinese training for antisubmarine warfare, and specifically to function as "targets" in ASW exercises. Development of an offensive submarine capability was not intended and US officials took some pains to point this out in congressional hearings, for despite its support for the Republic of China, the United States has been careful not to equip that country's armed forces with so-called "offensive weapons" ' (ibid., p. 236).

8. See Chi-wu Wang, 'Taiwan's Defense Policy in the Context of her Economic Development', unpublished paper prepared for the Conference on Security and Development in the Indo-Pacific, Fletcher School of Law and Diplomacy, Boston, Mass. (24–6 April 1976).

9. One example cited is that of microchip technology, which has applications to automotive and medical electronics and digital consumer products, as well as to proximity fuzes for anti-aircraft shells and other military equipment.

10. Chi-wu Wang, 'Taiwan's Defense Policy', op. cit., p. 17.

11. 'Chinese (Taiwan) Summary', in *Defense Market Survey, Intelligence Report: South American/Australasia* (Greenwich: DMS, Inc., 1981) p. 7.

12. US Department of State, from US Embassy, Taipei, cable no. 07721, 19 November 1978, p. 1.

13. Social Action Coordinating Committee, 'Brief on the Inertial Technology Training Program at MIT', (Cambridge, Mass.: April 1976) p. 6.

14. *Defense Market Survey*, op. cit., p. 3.

15. *Defense Market Survey*, op. cit., p. 7.

16. Interviews in Taiwan, 1982.

17. Chi-wu Wang, 'Taiwan's Defense Policy', op. cit., p. 14; US Department of State, cable no. 07721, op. cit., p. 1. Exports in Taiwan are limited as a result of deliberate policy (the government will not sell, at least openly, to any country that is not pro-American), by the nature of exports, by the relatively higher cost of production for all but the more simple munitions, and by the limitations imposed by US licensing agreements when the items in question contain US technical inputs or know-how. It is thought by some, however, that Taiwan would have few reservations about exporting to any non-communist country. Exports of small munitions and ammunition are thought to have gone to Latin America, South-east Asia, the Middle East and a few African countries (ibid., p. 2).

18. High-precision machinery production has been increasing since the late 1970s in cooperation with American firms, under the auspices of Taiwan's Industrial Technology Research Institute, which serves as a major center of military-industrial research in Taiwan. See Jeffrey M. Lenorowitz, 'Taiwan Technology Outlook Bright', *Aviation Week and Space Technology* (11 June 1979) 145.

19. Interviews conducted in Taiwan, 1982.

20. Chi-wu Wang, 'Taiwan's Defense Policy', op. cit., p. 18.

21. US State Department, cable no. 07721, op. cit.

22. Interviews at the American Institute on Taiwan (AIT). Strict review procedures are applied to requests from Taiwan for technical training in the US to determine that they involve civilian programs. Nevertheless, these are areas in which civilian training still can be 'dual-use', with spin-offs to defense-related activity.

23. Interview in Taiwan, March 1982.

24. Lenorowitz, op. cit., p. 145.

25. The XC-2 and XAT-3 programs may have been cancelled recently, according to US industry sources. See *Defense Market Survey*, op. cit., p. 8. Were the XAT-3 to be produced, it would be utilized to replace Taiwan's aging Lockheed T-33 trainers; the XC-2 was intended to supplement the air force's C-119s. Production of the KA-3 began in 1980 and was touted by the Taiwanese as 'a decisive step toward self-reliance'. Its engine is produced in Taiwan under foreign license. See *Defense et diplomatie*, 4, 32 (10 September 1979).

26. Another complement of F-5Es has been approved by the Reagan administration as a substitute for the F-5Gs requested by Taiwan and denied. No schedule of deliveries has been set for these additional aircraft at the time of writing.

27. Others argue that the F-8 is designed as a high-speed interceptor useful for defense against incoming bombers and not as an air superiority fighter that could give the PRC decisive control of airspace over Taiwan and the Taiwan Straits. See Robert G. Sutter and William de B. Mills, 'Fighter Aircraft Sales in Taiwan: US Policy', Congressional Research Service, issue brief no. IB81157 (20 January 1982) p. 4.

28. *Aviation Week and Space Technology*, (9 June 1980) 18; Chi-wu Wang 'Military Preparedness and Security Needs', op. cit., p. 657.

29. Northrop Corporation Presentation Document, 'Fighter Coproduction

and Security Assistance', NB81-95, Northrop Corporation Aircraft Division (Hawthorne, California, May 1981). Also, US Congress, Senate, Committee on Appropriations, *Foreign Assistance and Related Program Appropriations, Fiscal Year 1975, Hearings before a Subcommittee of the Committee on Appropriations*, 93rd Congress, 2nd Session, July 1974.

30. Chi-wu Wang, 'Taiwan's Defense Policy', op. cit., app.

31. Northrop Presentation Document, 1981, op. cit. This document, designed to allay the concerns of Congress and the public concerning the effect of coproduction on the US defense base and economy, states in bold face that 'F-5 technology transferred does not equip Taiwan to compete in the high performance fighter field'. This receives less emphasis, one assumes, in discussions with the Republic of China. However, it has been a major complaint of some Taiwanese defense analysts that the F-5 program, given that it did not include coproduction of the power plant, was little more than a coassembly scheme. Had the F-X been approved for coproduction, it is certain that the Republic of China would have pressed for much greater technical participation. See Chi-wu Wang, 'Military Preparedness and Security Needs', op. cit., p. 660. It should be noted, however, that over the course of the development of an aircraft industry in Taiwan, the AIDC has developed a complete wind tunnel for aerodynamic testing capabilities (from interviews in Taiwan, March 1982).

32. In 1975, a contract between the National Taiwan University and MIT was signed which provided for a two-year program to train fifteen students in the development, design and testing of inertial navigation and guidance systems. The program consisted of two parts: an academic program under the aegis of MIT's Center for Advanced Engineering and a 'hands-on' training program at the Charles Stark Draper Laboratories, an R&D corporation that specializes in inertial guidance systems for military application. The latter program was terminated by the State Department in 1975. According to the group of graduate students who focused attention on this program, Draper's role continued in another form as Draper engineers simply went to MIT to lecture in 'laboratory' settings. Ultimately, the State Department determined that the reasons given for the program – seeking low-cost, commerical technology for Taiwan – were spurious and that the program as a whole would require a US munitions license, which was denied; the students were sent home and their notes embargoed. Allegations by some observers suggested further that the program was a central part of Taiwan's nuclear development program. See Social Action Coordinating Committee, 'Brief on the Inertial Technology Training Program at MIT', unpublished (Cambridge, Mass., April 1976).

Taiwanese defense specialist Chi-wu Wang, Director of International Programs at the (Taiwanese) National Science Council displayed a candor rare in such matters when he acknowledged: 'It is no secret that air-to-air missiles and ground-to-air missiles are under development in Taiwan to improve its air defense network'. There are other reports in the open literature that Taiwan has actually developed a long range surface-to-surface missile. Professor James Hsiung, for instance, in a statement presented to a workshop sponsored by the Senate Foreign Relations

Committee in March 1980, stated categorically that Taiwan had devoted R&D resources to the development of an SSM with a 960 kilometer range, capable of reaching major Chinese population and industrial centers such as Canton, Foochow, Shanghai and Nanking. See Hsiung, 'The Security of Taiwan and US Policy', in US Congress, Senate, Committee on Foreign Relations, *Taiwan: One Year After United States–China Normalization* (Washington, DC: US Government Printing Office, 1980).

33. This is consistent with the perceived leverage afforded states, like Taiwan, thought to have a latent nuclear weapons capability.

34. *Defense Market Survey*, op. cit., p. 8.

35. Interviews in Taiwan, March 1982.

36. Like the Gabriel, the Hsiung Feng is reported to have a range of 1–26 nautical miles; *Defense Market Survey*, op. cit., p. 8.

37. Wang, 'Military Preparedness and Security Needs', op. cit., p. 659.

38. Taiwan press report from Foreign Broadcast Information Service, September 1979.

39. Durch, op. cit., p. 246.

40. Durch, op. cit., p. 247.

41. Ibid.

42. 'Shipping in Asia', *Far Eastern Economic Review* (28 February 1975) p. 9.

43. Interview at AIT, Taiwan, March 1982.

44. Ibid.

45. For a full discussion of the Republic of China ASW capabilities, see the detailed discussion in Durch, op. cit., pp. 234–6.

46. *Defense Market Survey,* op. cit., p. 9.

47. Wang, 'Military Preparedness and Security Needs', op. cit., app.

48. *Defense Market Survey*, op. cit., p. 12. The decision to procure the two submarines is thought to have been undertaken largely as a political expedient rather than on the basis of careful planning. It, in turn, precipitated expensive revision of planned defense allocations, engendering serious external controversy.

49. An exception to this is the recent development of a design for an attack boat 'with twice the fire power of the Tzu Chiang missile boat commissioned in 1980'. This was reported in the English language *United Daily News*, Taipei (15 March 1982). This attack boat reportedly will be equipped with Hsiung Feng SSMs and two rapid-firing guns. It was developed by China Shipbuilding and 'Taiwan National University', probably CIST.

50. US State Department, cable no. 07721, op. cit.

51. Interviews in Taiwan, March 1982.

52. *Defense Market Survey*, op. cit.

53. Adaptation of Rockeye in an improved anti-tank munition is thought to have played a prominent role in the 1982 Israel attack on Lebanon.

54. Interviews at AIT, Taiwan.

55. Ibid.

56. Among the incentives offered to Korean firms to engage in defense production were immunity from taxes on raw materials needed for

production, access to low interest government loans, reduced tariffs for capital equipment importation, elimination of taxes on profits, and a guaranteed 10 per cent profit on defense output (interviews in Seoul, 1982). The ensuing discussion draws on interviews conducted with South Korean defense planners, the Joint US Military Advisory Group in Korea (JUSMAG), US Embassy officials in Seoul, and officials of the US State and Defense Departments.

57. One Korean government official in the Korean Development Institute insisted that demand projection studies were conducted but turned out to be inaccurate. A more prevalent view is simply that no requirement studies were done in advance of the crash manufacturing program, and that the Park régime encouraged many industries to duplicate each other because it was not clear which, if any, firm would succeed.

58. For more extensive discussion, see Chapter 5 and Richard M. Curasi, 'Neither Puppet nor Pariah Be: the Development of the South Korean Defense Industry as an Indicator of the Quest for Independence, Self-reliance, and Sovereignty', unpublished paper prepared for the Naval Post-graduate School, Monterey, California (September 1979).

59. Quoted in US Congress, Senate, Committee on Foreign Relations, *US Troop Withdrawal From the Republic of Korea* (Washington, DC: US Government Printing Office, July 1978) p. 52.

60. Ibid., p. 51.

61. *Aviation Week and Space Technology* (3 May 1982) p. 18.

62. Interviews at the Korean Development Institute, Seoul, March 1982.

63. Interviews at US Embassy, Seoul, March 1982.

64. Interview with Korean industrialist, 1982.

65. Interviews at US Embassy, Seoul, March 1982.

66. Representative of Poong-San Metal Manufacturing Corporation (PMC), Seoul, 1982.

67. Alain Cass, 'Start Made on Changing Structure', *Financial Times*, 19 August 1983.

68. Meetings at KIDA and with JUSMAG, Seoul, 1982.

69. Political and economic problems are compounded by the concentration of particular industries in conglomerates around the country. By deliberate government design, firms were encouraged to locate in quasirural areas in need of development. This suggests that plant closures could have a severe economic impact in areas where the local economy is highly dependent on employment in the firm.

70. The major sources for this section were interviews with the US and Korean government and industry representatives in Seoul. Other documentation includes 'Defense Industry in Korea', a paper produced for my use by the Poong-San Metal Manufacturing Company in Seoul, and a memorandum of conversation with Dr Hwang Dong Joon, Korean Institute for Defense Analysis, produced by JUSMAG-Korea (reporting a meeting of Dr Hwang, a JUSMAG representative, and myself, March 1982).

71. Press reports of the April 1982 visit to Korea by Secretary of Defense Caspar Weinberger indicate that no changes in US export restrictions are imminent. Executive Branch latitude in this regard, even if it were

intended, is circumscribed by congressional and industrial pressures against any liberalization of coproduced exports, which are thought to harm both US commercial interests and the US defense industrial base. Even though a number of items in question are no longer produced in the United States, this issue is highly contentious in the Congress. See Chapter 2 for discussion of recent Congressional initiatives to adopt even more stringent protectionist measures for US technology, and Richard Halloran, 'Weinberger says US will Maintain Curbs on Seoul's Sale of Arms', *New York Times*, 1 April 1982, p. 5.

The extent to which Korea has officially tried to underplay the heavy dependence of its production programs on the United States was exemplified in a film shown to me at KIDA in March. The film was a crashingly enthusiastic interpretation of the defense production capabilities of Korean industries. It managed completely to avoid reference to US assistance and components. One notable example was the so-called 'NH-K' missile (Nike Hercules-Korea), the product of a cooperative venture with the United States that began initially as a maintenance program in 1972. No reference to any US role was made in the description of this major success story. For further discussion, see *Wall Street Journal*, 10 January 1978, p. 1, and 'South Korea's Arms Industry: Boom Boom', *Economist*, vol. 269, no. 7057 (2 December 1978) p. 36.

72. Interview at KIDA, Seoul, March 1982.
73. 'Administration Seeks Korean Aid Hike', *Aviation Week and Space Technology* (3 May 1982) p. 18, and *Defense Market Survey*, op. cit., p. 10.
74. Interviews in Seoul, March 1982, and *Defense Market Survey*, op. cit., pp. 19–20. The present cost of the F-16 program, including support equipment, spare parts, training and technical aid is estimated to be over $900 million. Although deliveries will not begin until 1986, the Koreans have already begun paying, using a portion of fiscal year 1982 FMS credits.
75. *Defense Market Survey*, op. cit., p. 10.
76. Estimates of the number of helicopters built vary according to the source. The range is from 66 to 75, not counting the additional 48 units referred to above. See *Defense Market Survey*, op. cit., p. 10 and 'Koreans see New Military Capacity', *Aviation Week and Space Technology* (22 October 1979) p. 62.
77. Ibid., p. 63.
78. Part of the agreement included the purchase of 25 units equipped with TOW missiles which were not included in the coassembly program. It is reported that the Koreans did want to produce TOW missile-equipped helicopters themselves, and may still hope to do so. See ibid., p. 63.
79. Ibid., p. 63, and US Congress, Senate, Committee on Foreign Relations, *US Troop Withdrawal*, op. cit., p. 52.
80. '*Koreans See New Military Capacity*', op. cit., p. 63. Training opportunities arising from Korea's aircraft industry are discussed in a later section and in Chapter 4.
81. US Congress, Senate, Committee on Foreign Relations, *US Troop Withdrawal*, op. cit., p. 14.

82. From interviews in the US State Department, 1981, and Curasi, 'Neither Puppet nor Pariah Be', op. cit., p. 14.

83. Interviews in the US State Department and with industry officials, 1981. A major incentive for trying to develop such systems seems to have been Park's perception – and resentment – of superior missile capabilities in the North. South Korea's development efforts would have been directed at countering the threat posed by the North Korean deployment of FROG-7 SSMs and at rectifying perceived disparities between South Korean and North Korean missile production capabilities.

84. See note 71.

85. *Defense Market Survey*, op. cit., p. 11. A proposal was offered in 1977 to sell Korea kits to upgrade its inventory of Hawk missiles.

86. *Ground Defense/Armies and Weapons* (November 1979) p. 5.

87. Ibid. While missile production has, from the outset, been an important aspect of Korean defense industrialization, the acceleration of cooperative programs with the United States in the 1970s may have been due to the pressures being placed by independent Korean programs on official US interests. Dissuading the Koreans, or any recipient, from engaging in programs considered inimical to US interests requires that the military/ psychological concerns of the recipient in a particular program be met through other channels. Deliveries of additional weapons (and/or more cooperative programs) seem to be the 'treatment of choice'.

88. US General Accounting Office, *Coproduction Programs*, op. cit., p. 8.

89. *International Defense Review* (November 1983) p. 1537.

90. Defense Market Survey, op. cit., p. 11, and Korean Overseas Information Service, Ministry of Culture and Information, *A Handbook of Korea* (Seoul: Samhwa Printing Co., 1979) p. 452.

91. *Defense Market Survey*, op. cit., p. 12.

92. US Congress, Senate, Committee on Foreign Relations, *U.S. Troop Withdrawal*, op. cit., p. 53.

93. *Military Technology*, August 1983, p. 92; *International Defense Review*, November 1983, pp. 1535–6; *Jane's Defence Weekly*, 28 January 1984, p. 104.

94. *Defense Market Survey*, op. cit., p. 12.

95. US Congress, Senate, Committee on Foreign Relations, *US Troop Withdrawal*, op. cit., p. 14.

96. Ibid., p. 56.

97. Interviews at US Embassy, Seoul, 1982.

98. PMC has three principal plants. Two are engaged exclusively in defense work (although diversification into sporting ammunition is beginning). The other is a brass plant, and produces, among other things, coin blanks for export. The latter is the only plant which is more than 50 per cent active, and it too has suffered cutbacks.

99. Interviews with PMC directors, March 1982.

100. Economic Planning Board, Republic of Korea, 'A Summary Draft of the Fifth Five Year Economic and Social Development Plan 1982–1986' (September 1981).

101. See *Defense Market Survey*, op. cit., p. 49.

102. Interviews in Seoul, 1982.

103. Chi-wu Wang, 'Military Preparedness and Security Needs', op. cit., p. 659.
104. For illustrative discussions, see Young-sun Ha, 'Nuclearization of Small States and World Order: the Case of Korea', *Asian Survey*, xi (November 1978) 1134–51, and Franklin B. Weinstein and Kamiya Fuji, *The Security of Korea: US and Japanese Perspectives on the 1980s* (Boulder, Col.: Westview Press, 1980) ch. 4.

Chapter Four: Modernization in the Garrison State

1. An exception to this general rule is the climatic conditions that hurt agriculture production in the later 1970s.
2. In order to stimulate and stabilize the economy, the Taiwan government in late 1981 proposed targets for reductions in energy consumption, including a reduction in external dependency on oil from 71 per cent to 49 per cent in ten years. See Samuel P. S. Ho, 'South Korea and Taiwan: Development Prospects and Problems in the 1980s', *Asian Survey*, xxi, 12 (December 1981) 1194, and American Institute in Taiwan, 'Foreign Economic Trends and their Implications for the United States', draft (December 1981) p. 3. Reduction in oil consumption is based on projections that Taiwan can fulfill 28 per cent of its energy needs in coal and 14 per cent in nuclear power by 1989. These are optimistic projections by most accounts. The incentive, however, is great. Taiwan's oil import expenditures in 1980 increased by 89 per cent over the previous year and accounted for 21 per cent of total imports.
3. Council for Economic Planning and Development, *Ten Year Economic Development Plan for Taiwan, Republic of China (1980–1989)* (Taipei, March 1980).
4. Two separate sets of stabilization measures were introduced, one before Park's assassination and one after. In the prior case, some argue that the economic dislocations of the time – including massive consumer price increases and shortages of consumer goods, – followed by stringent fiscal and monetary policies whose rapid introduction caused bankruptcies in some industries – created the conditions for unrest and insurrection. See Chong-Sik Lee, 'South Korea, 1979: Confrontation, Assassination and Transition', *Asian Survey*, xx, 1 (January 1980) 63–76.
5. Chong-Sik Lee, 'South Korea in 1980: the Emergence of a New Authoritarian Order', *Asian Survey*, xxi, 1 (January 1981) 139; Samuel P. S. Ho, op. cit., p. 1188.
6. At the end of 1979, 259 of the 334 major enterprises in Korea owed $5.66 billion to foreign investors. See Chong-Sik Lee, 'South Korea in 1980', op. cit., p. 140.
7. Samuel P. S. Ho, op. cit., pp. 1166–7.
8. Nena Vreeland et al., *Area Handbook for South Korea* (Washington, DC: US Government Printing Office, 1975) p. 215.
9. For a detailed discussion of the economic development of Taiwan and South Korea prior to 1950, see, for instance, Samuel P. S. Ho, 'Economic

Development of Taiwan, 1860–1970', manuscript (New Haven, 1977); George W. Barclay, *Colonial Development and Population in Taiwan* (Princeton University Press, 1954); and Walter Galenson (ed.), *Economic Growth and Structural Change in Taiwan* (Ithaca & London: Cornell University Press, 1979) chs 1–3. For Korea, see Hochin Choi, 'The Process of Industrial Modernization in Korea: the Latter Part of the Chosun Dynasty through the 1960s', *Journal of Social Sciences and Humanities*, 26 (June 1967) 1–33; W. D. Reeve, *The Republic of Korea: a Political and Economic Study* (London: Oxford University Press, 1963); and Paul W. Kuznets, *Economic Growth and Structure in the Republic of Korea* (New Haven & London: Yale University Press, 1977) ch. 1.

10. Prior to the influx of immigrants to Taiwan from mainland China, Taiwan had a population of just six million. Total production was particularly low in 1946 and 1947, amounting to just over half of the net domestic product of 1937. Dislocation of the economy was particularly evident in mining, manufacturing and construction, reflecting both the dangers of the war and the social upheaval which resulted from Japanese evacuation and the legacy of colonialism. See Simon Kuznets, 'Growth and Structural Shifts', in Walter Galenson, op. cit., pp. 27–36.

11. Paul Kuznets, op. cit., pp. 28–9.

12. Ian M. D. Little, 'An Economic Reconnaissance', in Galenson, op. cit., p. 450.

13. Ibid., p. 461.

14. Both countries pursued a strategy of import substitution during the 1950s, a strategy based on the restriction of imports which could compete with fledgling production structures in the indigenous economy. This was employed largely as protection against non-durable consumer goods industries and intermediate goods required for their manufacture. This strategy required policies ranging from controls on foreign exchange to licensing requirements for imports, and was usually accompanied by deficit financing, inflation and an increasingly overvalued currency. This strategy is discussed in more detail in a later section of this chapter. The quote is from Maurice Scott, 'Foreign Trade', in Galenson, op. cit., p. 377.

15. The régime of Syngham Rhee was a period of serious economic and political paralysis, marred by political corruption and despotic rule. Among the more salient types of social dislocations was the disaffection of the military with the ruling régime, at least a partial outgrowth of the Rhee-inspired political environment that was rife with factionalism, ideological rivalries and petty intrigues. The military during this period was in a state of constant dissent, which sowed the seeds for subsequent insurrection.

16. See Anne O. Krueger, *The Development Role of the Foreign Sector and Aid* (Cambridge, Mass.: Harvard University Press, 1979) pp. 1970–81 and Edward S. Mason *et al.*, *The Economic and Social Modernization of the Republic of Korea* (Cambridge, Mass. & London: Harvard University Press, 1980) ch. 1. One example of the Rhee régime's resistance to investment assistance for the infrastructural development of the South is the area of power production, for the implementation of which projections of future demand for electricity were required. Rhee refused to have these

conducted on the basis of assumptions that there would be no supplies
from the North since this would have undermined the government's
position on the need for reunification prior to further development of the
South.

17. Maurice Scott, op. cit., p. 369. Estimates vary widely in the literature, in
part because aid continued beyond 1964 in the form of yet undelivered
commitments, which skews any calculation of an average. In addition,
estimates vary according to whether or not they include allocations for
military support; this figure does not. For the military figure, see Ian M.
D. Little, op. cit., p. 458. American economic assistance to Taiwan during
this period was also closely linked to protecting domestic stability in an
economy that was so fragile that the consumer price index rose about
ten-fold a year in the period 1946–9, about 500 per cent in 1949–50, and
continued at a rate of 80–100 per cent a year until the early 1950s.
Inflation was closely linked to the unusually high military burden being
supported by Taiwan, with the military representing about 20 per cent of
the civilian labor force. In the absence of American assistance to support
the military and to sustain the economy through its period of postwar
dislocation (marked by shortages of commodities and trade deficits, in
particular), it is unlikely that political stability could have been assured
during this period. Aside from the political considerations, aid may have
played an important role in influencing the choice of economic policies
during this time, but this is a source of contention among economists. For
competing views on this subject, see Maurice Scott, op. cit., pp. 371–4,
which presents his and summaries of others' arguments, and Neil H.
Jacoby, *US Aid to Taiwan* (New York: Praeger, 1966) chs 2, 3. For the
role of aid on the development of Korea during this period, see Anne O.
Krueger, op. cit., pp. 23–52.

18. For a full discussion of monetary and fiscal policy during this period, see
for Taiwan, Erik Lundberg, 'Fiscal and Monetary Policies', in Galenson,
op. cit., ch. 4, and for Korea, Krueger, op. cit., ch. 2, and Paul Kuznets,
op. cit., chs 2, 3.

19. The literature on the domestic political structures of Korea and Taiwan
is highly controversial and reflects serious ideological schisms. For
representative discussions of Taiwan, see, for example, Victor H. Li (ed.),
The Future of Taiwan: a Difference of Opinion (White Plains, N. Y.:
M. E. Sharpe, 1980); John Franklin Copper, 'Political Development in
Taiwan', in Hungdah Chiu (ed.), *China and the Taiwan Issue* (New York:
Praeger, 1979) pp. 37–77; Ralph Clough, *Island China* (Cambridge, Mass.:
Harvard University Press, 1978) chs 2, 4; and Jurgen Domes, 'Political
Differentiation in Taiwan: Group Formation within the Ruling Party and
the Opposition Circles 1979–1980', *Asian Survey*, xxi, 10 (October 1981).
For Korea, see Se-Jin Kim, *The Politics of Military Revolution in Korea*
(Chapel Hill: University of North Carolina Press, 1971); Harold Hakwon
Sunoo, *America's Dilemma in Asia: the Case of South Korea* (Chicago:
Nelson Hall, 1979); Gregory Henderson, 'Korea', in Henderson *et al.*,
Divided Nations in a Divided World (New York: David McKay, 1975) ch.
2; and Franklin B. Weinstein and Fuji Kamiya, op. cit.

20. The linkages among industry and government in the defense production

area is not a subject which is discussed in the countries in question as it is considered highly sensitive. The analysis contained here is based on inferences drawn from interviews conducted in Seoul and Taipei in 1982, and from interviews with and published articles by American observers. A major constraint is the absence of data on the distribution of personnel in defense activities. To my knowledge, this is not a subject that has ever been discussed directly, either in the United States or in the countries in question.

21. There are also 210,000 aborigines in Taiwan who are of Malay origin. This group has almost no political voice in the country. See Yung Wei, 'Political Development in the Republic of China on Taiwan', in Yung-Hwan Jo (ed.), *Taiwan's Future?* (Tempe: Arizona State University, 1974) p. 25, and Hung-mao Tien, 'Taiwan in Transition: Prospects for Socio-Political Change', *China Quarterly*, 64 (December 1975) 626. Figures for the proportion of Taiwanese to mainlanders are not provided in official sources in Taiwan, as this is part of the effort to deflect attention from the divisions in the population.

22. In November 1981, the Central Standing Committee of the KMT appointed nine new leaders to the Executive Yuan, five of whom are native Taiwanese. This was the first cabinet change since 1978, and reflects the gradual integration of Taiwanese in the upper echelons of policy-making. See *Asia Report*, 2, 12 (December 1981) 2–3. All major decisions are ultimately the province of the Central Standing Committee of the KMT, in which Chiang Ching-kuo and the chief economic and military planners are members. This body supersedes in authority all other government branches and agencies. The KMT as an organization actually has relatively little influence over national military, economic or foreign policy decisions. See Clough, *Island China*, op. cit., pp. 49–54. Taiwanese domination of the commercial sector has led to a pattern of intense discrimination in employment against mainlanders, who must seek employment in the government, public sector projects, or the marginal number of businesses run by mainlanders. This could serve as a significant source of political instability if economic difficulties lead to more pro-nounced discrimination against mainland candidates when the public sector can no longer readily assimilate them. See Hungdah Chiu, 'The Future of Political Stability in Taiwan', in US Congress, Senate, Committee on Foreign Relations, *Taiwan: One Year after United States–China Normalization* (Washington, DC: US Government Printing Office, 1980) p. 41.

23. The ideological divisions among Taiwan's élites is a subject of vast complexity which exceeds this analysis. As a simple historical summary, it should be noted that the advent of mainlander control over the government of Taiwan in the late 1940s began with the division among advocates of what became the reigning ideology, based on the pledge to restore the KMT as the only legitimate government of all of China, and those who were more inclined towards separatism for Taiwan. Restoration of KMT rule over the mainland became the basis of perpetuating a mainlander-dominated government, since a government reflecting the majority of people on Taiwan would imply acceptance of the status quo,

a separate state. The 'one-China' ideology is still the grounds on which KMT legitimacy is based. Although there has been an increasing ascendancy of political moderates in Taiwan who play down the 'one-China' mythology, no politicians in Taiwan (with the exception of the radical Taiwan Independence Movement) refer directly to Taiwanese independence. Rather, they are careful to phrase their statements in terms of the desirability of expanding the self-reliance and self-determination commensurate with the status of a modern power. It should be added that the interest of mainlanders to control the government of Taiwan has an economic motive. Mainlanders arriving in Taiwan in the late 1940s owned neither the land nor businesses and thus had no other way of making a living. For a full discussion of this period, see for example, Ralph Clough, *Island China*, op. cit., chs 1–6; Hungdah Chiu (ed.), *China and the Taiwan Issue*, op. cit., chs 2, 3; and Nathaniel B. Thayer, 'China: the Formosa Question', in Henderson *et al.*, *Divided Nation in a Divided World*, op. cit., ch. 3.

24. Domes, op. cit., pp. 1024–6.
25. Korean social life is strongly influenced by religious forces, emanating mostly from Confucianism adapted from China and Christianity adapted from the West, with only a few other strains, such as Buddhism, having a significant influence. For a detailed discussion of the religious and cultural influences on the structure of sociopolitical development in Korea, see Se-Jin Kim and Chi-Won Kang, *Korea: a Nation in Transition* (Seoul: Research Center for Peace and Unification, 1978) chs 4, 8, 9, 10; Gregory Henderson, *Korea: the Politics of the Vortex* (Cambridge, Mass.: Harvard University Press, 1968) ch. 2; and Harold Hakwon Sunoo, op. cit., ch. 8. The legacies of Confucianism that are most often referred to as strong influences on Korean political life are the centralization of authority around a ruling figure, a strong respect for family and kinship as the source of sociopolitical identity, and a presupposition that there should be a separation between the rulers and the ruled. In skeletal form, this suggests that primal forces actually continue to determine the centralization of authority in Korea, whose official justification now derives from modern interpretations of the need for strong government and public control. An interesting interpretation of Korean class structure as derivative of Confucianism likens the Confucian 'scholar-gentry' élitism with the structure of modern Korean state bureaucracy; both serve as 'the major bulwark of the ruling class against any changes or reforms of the political system'. See Sunoo, op. cit., ch. 8.
26. The centralization of authority in the early 1970s has been attributed to several causes, depending on the perspective of the observer. See Vreeland *et al.*, op. cit., pp. 171–82; Chong-Sik Lee, 'South Korea 1979', op. cit.; and Joungwon A. Kim, *Divided Korea: the Politics of Development 1945–1972* (Cambridge, Mass: Harvard University Press, 1975) ch. 7.
27. The six points were: 'The government would henceforth give top priority to national security measures; social unrest endangering national security would not be tolerated and all elements of such unrest would be stamped out; "irresponsible" discussion of national security would be banned; every citizen would have to adopt for himself a new sense of values

consistent with the needs of national security; and every citizen would have to sacrifice some of the freedom he was enjoying.' See Vreeland, *et al.*, op. cit., p. 175. The effort to extend Park's tenure beyond 1971 was officially justified in part on the assumption that North Korea had active plans to seize the South by the early 1970s. The new constitutional system granted Park the right to succeed himself indefinitely, if elected by a special electoral college, one-third of which was nominated by Park himself. Throughout the 1970s, Park utilized the emergency powers given him under the new constitution to stymie opposition. See Chong-Sik Lee, 'South Korea 1979', op. cit., pp. 64–5, and Sunoo, op. cit., epilogue.

28. See Chi-wu Wang, 'Taiwan's Defense Policy in the Context of her Economic Development', unpublished paper prepared for the Conference on Security and Development in the Indo-Pacific, Fletcher School of Law and Diplomacy, Boston, Mass. (24–6 April) p. 17.

29. Korea did not entirely neglect the development of a scientific infrastructure to assist in the development of weapons. The Agency for Defense Development, for instance, was instituted in the late 1960s, while much attention was given to nuclear energy programs through institutes such as the Korea Nuclear Development Corporation, the Korea Institute of Science and Technology and, most importantly, the Korea Atomic Energy Research Institute. These R&D efforts were almost entirely under government aegis, however, with only marginal linkages to the private sector. Thus, the effort to promote defense industrialization created a government-controlled scientific community and did not produce links with, or encourage, Korea's private scientific community. No effort was made to include Korea's university community, which remains a major source of political opposition and unrest today.

30. In the early stages of the Park administration, the military officers who had helped to engineer the coup remained in government as influential administrators. In later years, many joined the private sector but retained their close ties to the Park régime. By the 1970s, they were competing for prestige at the top of the hierarchy with civilian technocrats – engineers, economists and management experts – whose planning expertise was increasingly in demand. See Edward A. Olsen, ' "Korea, Inc.": the Political Impact of Park Chung-Hee's Economic Miracle', *Orbis* (Spring 1980) 74. Generals in the armed forces continued to be an integral part of the overall power structure, however, in spite of official strictures on their political stature which stressed that the armed forces were 'neutral' and non-partisan. See Vreeland *et al.*, op. cit., pp. 181–2 for more detailed discussion of the role of the armed forces in Korean political development.

31. See Gustav Ranis, in Galenson, op. cit., p. 215. The institutes offering assistance included, for instance, the China Productivity and Trade Center, the Food Industry Research and Development Institute, the China Development Corporation, and the Industrial Development and Investment Center.

32. Ian M. D. Little, op. cit., p. 461. Aid also financed most of the imports necessary to keep the economy functioning, in light of the low export potential capable of earning necessary foreign exchange to finance imports.

For a full discussion of this period, see, for instance, Anne O. Krueger, op. cit., ch. 2.

33. For a full discussion, see Gustav Ranis, 'Industrial Development', in Galenson, op. cit., ch. 3; Erik Thorbecke, 'Agricultural Development', in ibid., ch. 2; and Anthony Y. C. Koo, 'Agrarian Reform, Production and Employment in Taiwan', in International Labor Office, *Agrarian Reform and Employment* (Geneva, 1971).

34. Ian M. D. Little attributes this early approach to a traditional Chinese notion of 'developing agriculture by means of industry and fostering industry by virtue of agriculture', which he argues was crucial to the underpinnings of the economic 'take-off' period after 1960. See Little, op. cit., pp. 214–15.

35. For a full discussion of food and agrarian policy during this period, see Young Whan Khil and Dong Suh Bark, 'Food Policies in a Rapidly Developing Country: the Case of South Korea, 1960–1978', *Journal of Developing Areas*, 16 (October 1981) 47–70.

36. Samuel P. S. Ho, 'South Korea and Taiwan: Development Prospects and Problems in the 1980s', *Asian Survey*, xxi, 12 (December 1981) 1176. There is a voluminous body of literature on the economic transformation of South Korea and Taiwan during this period, examples of which are cited below. For the purpose of this study, only the major highlights of economic growth are presented to serve as the basis for analysis of the way in which development strategy and defense modernization began to interact as agents of development during the 1970s. For full discussion of this period, for Taiwan, see Ching-yuan Lin, *Industrialization in Taiwan, 1946–1972* (New York: Praeger, 1973) chs 5–7; Walter Galenson (ed.), *Economic Growth and Structural Change in Taiwan*, op. cit.; US Congress, Senate, Committee on Foreign Relations, *Taiwan: One Year After United States–China Normalization* (Washington, DC: US Government Printing Office, June 1980) pp. 60–117; and John C. H. Fei *et al.*, 'Growth With Equity: the Taiwan Case', World Bank report 0-19-52-0116-7 (1976). For Korea, see Anne O. Krueger, op. cit., chs 3–6; Edward S. Mason *et al.*, *The Economic and Social Modernization of the Republic of Korea* (Cambridge, Mass.: Harvard University Press, 1980); and Nena Vreeland *et al.*, op. cit., ch. 10.

37. See Little, op. cit., p. 475.

38. See Ranis, op. cit., pp. 221–2, 236–7.

39. Little, op. cit., pp. 482–3.

40. It is widely understood that the United States terminated aid to Taiwan because the US foreign assistance program needed a 'successful aid graduate'. Military and some food aid continued after 1965 and helped alleviate some of the potential dislocations that this policy change might have engendered. While the choice of Taiwan as guinea pig reflected US confidence in its economic capabilities, the Taiwanese did not take this measure as a compliment.

Calculated on the basis of gross domestic capital formulation, private foreign investment was quite marginal during the 1950s, but grew to over 11 per cent by 1971, which represented 19 per cent of industry overall and much higher percentages in individual industries, such as electronics, where foreign capital was most concentrated. See Economic Planning Council, *Taiwan Statistical Data Book* (Taipei, 1976); and Ranis, op. cit., p. 250.

41. See International Monetary Fund, *Balance of Payments Yearbook, 1964–1968* (Washington, DC: IMF, 1970).

42. Vreeland *et al.*, op cit., p. 313. A gross indicator of external dependence can be derived from the ratio of current payments deficits to domestic investment. Korea's ratio was an annual average of three-quarters up to 1963 and remained fairly high – one-third – for at least another decade. This compares to Taiwan, whose annual average ratio was zero between 1963 and 1973, after a much higher ratio (ranging between one-half and one–third) between 1951 and 1959. See Ian M. D. Little, op. cit., pp. 459–61.

43. Malcom H. Perkins and William Bolles, 'Security Assistance to South Korea: Assessment of Political, Economic and Military Issues from 1975 to 1979, Masters thesis, Air Force Institute of Technology, Air University (1979) p. 280. The United States contributed aid to Korea between 1950 and 1972 in the amount of about $5.5 billion. Korea also received loans from the Export-Import Bank in the amount of $391 million over the same time period, and received assistance from other international organizations in the amount of $759 million between 1956 and 1973.

44. A former AID official, noting the business opportunities being offered Koreans in Vietnam, referred to Vietnam as the 'El Dorado of Korea' – a place for rapid fortunes. See Perkins and Bolles, op. cit., p. 280.

45. Debt service payments increased from about $20 million in 1966 to over $308 million by 1972; the ratio of these payments to total export revenues increased by 6 per cent in 1966 to 18 per cent in 1972, in spite of large expansion of export sales; see Ian M. D. Little, op. cit., p. 489. The Korean government has, on occasion, tried to compensate for the lack of support derived from the agricultural sector by imposing harsh fiscal and monetary measures, but resources mobilized by decree have never been sufficient to keep pace with Korea's internal demand for investment resources. This is mosty because the government's policy of neglecting agricultural growth left this sector too weak to serve as a source of savings.

46. Vreeland *et al.*, op. cit., p. 350. The military, moreover, was used in a direct development fashion in civic action programs, which sent soldiers to assist rural communities to build roads and dams as well as to participate directly in agricultural activities. The most significant of these military-directed agricultural projects was the Reconstruction Village Program, begun after 1965, which provided for the relocation of rural populations to underdeveloped areas in the northern sector of South Korea, close to the Demilitarized Zone. This is a classic example of a defense/development link: the project was intended not only to make use of underutilized resources and to provide employment to subsistence farmers, but was a measure designed to provide a human 'early warning system' in the event of incursions from the North.

47. Vreeland *et al.*, op. cit., p. 318.

48. See Ranis, op. cit., p. 226 and Ching-yuan Lin, op. cit., pp. 107–8.

49. Ching-yuan Lin, op. cit., p. 107, citing Yu-chi Tao, *Chinese Customs and Tariffs: System and Practice* (Taipei, 1968) pp. 130–1. Taiwan did initiate a series of government plans to guide economic development. A series of six four-year plans, beginning in 1953, were produced by the government, the most significant of which was the Nineteen Point Program of Economic and Financial Reform (1961–4) which covered all the changes in trade policy and exchange rates discussed above. For the most part, regardless of the plans, actual economic change overshadowed the designs promoted by planning, and they seem to have had little influence over actual modernization policy. See Ian M. D. Little, op. cit., pp. 489–90.

50. Samuel P. S. Ho argues: 'Given the Korean government's predilection for promoting favored industries and enterprises, it is not surprising that a relatively small number of industrial firms have accounted for a large share of Korea's industrial production and exports and that many industries have oligopolistic structures.' See Samuel P. S. Ho, 'South Korea and Taiwan', op cit., p. 1195. See also Vreeland *et al.*, op. cit., pp. 318–20, and Paul Kuznets, op. cit., ch. 4.

51. See Chi-wu Wang, 'Taiwan's Defense Policy', op. cit., pp. 19–20.

52. Ship repair and maintenance did begin in the 1960s, serving as the basis for current shipbuilding capabilities; see Chapter 3.

53. Taiwan was able to disperse industries on the island in part because of topographical features which favored the West (fertile plains areas) and the South (port facilities), both of which had high concentrations of arable land and population. This strategy would not have been possible, however, without the infrastructural investments made in the 1950s and 1960s in roads and rail links, improving upon the network of transport facilities inherited from the Japanese. This network, moreover, was closely tied to the main ports, facilitating exports. See Ranis, op. cit., p. 222; also Ching-yuan Lin, op. cit., chs 6–7.

54. Although size of agricultural allocations have grown as the economy has developed, the ratio of investment in agriculture to total investment has declined precipitously, down from 30 per cent in 1961. To his credit, arguably, Park initiated a program in 1961–2 intended to assist poor farmers by replacing their private debts with low-interest government loans. This was a political expedient made necessary by Park's public pronouncements proclaiming the government's dedication to the welfare of the underprivileged rural population. He was then the Chairman of the Revolutionary Committee on National Reconstruction (sic). The program was discontinued very shortly after it was discovered that the farmers tended to use government loans 'for purposes other than farming'. See Young Whan Khil and Dong Suh Bark, op. cit., pp. 50–1.

55. Ibid., pp. 51–2. It must be considered that domestic food shortages and excessive dependency on outside sources for basic staples would be a serious problem in the event of military conflict, when food deliveries might be interrupted. This may add to Korea's security problems in the long run.

56. Electro-mechanical devices, such as microchips, received special attention for their potential application to both defense and civilian industry. It was a deliberate goal in 1976 to harmonize defense and economic

modernization objectives so that, where possible, defense advances would be directed into areas of maximum remuneration. See Wang, 'Taiwan's Defense Policy', op. cit., p. 15.

57. Some of these projects were begun before 1973. Publicizing them in this way (1973 was the first time that economic plans revealed specific projects) may have been aimed at maximizing public support for the government's plans; see Ian M. D. Little, op. cit., p. 450. Steel production and the two nuclear power plants which were included in this plan accounted for two-thirds of the estimated cost; see Prybyla, 74; see also Wang, 'Taiwan's Defense Policy', op cit., p. 15.

58. See Samuel P. S Ho, 'South Korea and Taiwan', op. cit., p. 1195.

59. See Chi-wu Wang, 'Taiwan's Defense Policy', op cit., p. 16. Wang argues that the quest for alternative energy sources is planned with defense considerations in mind. He gives geothermal energy as an example: 'Estimates of up to 4,000 megawatts have been made in Taiwan's northeast coast, over her entire Central Mountain Range and in the rift valley on the island's southeast. Thus geothermal energy may account for up to 20 per cent of Taiwan's energy supplies by the year 2000. The significance of this in terms of the ROC's defense is that geothermal power plants are likely to be built in small clusters and in remote mountain areas, and are therefore relatively safe from aerial bombardments' (ibid., p. 26). In addition, Taiwan has been diversifying its sources of petroleum by geographical area. While Kuwait and Saudi Arabia remain principal suppliers, the Chinese Petroleum Corporation has contacted Mexico for supplies; has started importing crude oil from Nigeria, Oman, Ecuador and Gabon; has entered into contract negotiations with Indonesia to build a naphtha plant in that country; and is planning to invest in a petroleum refinery in Hawaii. Taiwan also receives coal from the United States, Canada, Australia and South Africa and uranium from the US and South Africa; and it is involved in a venture to explore for uranium in Paraguay. See American Institute in Taiwan, 'Foreign Economic Trends and their Implications for the United States' (Taipei, December 1981).

60. The ambitious nature of Park's development plans has always been subject to criticism, both domestic and foreign. The relative success of these plans in some areas, however, beginning as early as the first five-year plan, has consistently exceeded expectations in many ways. For instance, annual per capita income rose from $96 to $131 between 1962 and 1966, and the revenues realized from exports were nearly double the planning goal. See Kuznets, op. cit., p. 46. Nevertheless, the plan was not an overall success. Policies to promote rapid growth created excessive expansion of credit, a declining level of foreign exchange holdings, and reductions in real wage rates in both manufacturing and agriculture. These adverse effects stemmed in part from inflation, which increased the annual average price of consumer goods by 17 per cent. Real wages in manufacturing fell by 1966 to 90 per cent of the 1960 level. See Vreeland, *et al.*, op. cit., p. 228.

61. Samuel P. S. Ho, 'South Korea and Taiwan', op. cit., p. 1181.

62. See, for instance, World Bank, 'Report and Recommendations of the President of the International Bank for Reconstruction and Development to the Executive directors on a Structural Adjustment Loan to the

Republic of Korea', unpublished, report no. P-3156-KO (17 November 1981) p. 20.

63. A conservative estimate of the premium paid by countries to coproduce rather than purchase weapon systems can be derived from the costs incurred by European producers from participation in the F-16 program. Even in what might be considered an 'ideal' environment – solid technical and managerial infrastructure, a skilled work force, and other industrial benefits lacking in Korea – the cost of production over outright purchase was as much as 15 per cent. See Carl Groth, 'The Economics of Weapons Coproduction', paper presented to the International Studies Association Conference, Los Angeles (20 March 1980).

64. Edward Olsen suggests that companies such as Hyundai and Daewoo – 'superconglomerates' as he calls them – receive government support in direct proportion to the regularity with which they funnel private funds to government officials. See Olsen, op, cit., pp. 69–84. Korea's reliance on external sources for development financing was extremely high. By some estimates, Korea had acquired foreign loans in excess of $1.9 billion by 1969, and over $3 billion by 1971. Many of these loans were acquired without regard for the feasibility of the industrial projects to be undertaken, or of the ability of Korean enterprises to repay. External indebtedness has continued to be a problem in Korea, given the government policy of stressing economic expansion over economic stability. For more detailed discussion, see Joungwon A. Kim, op. cit., pp. 274–96. The World Bank estimates that of the investment funds required in 1978, over 75 per cent was provided through debt. Thus, bankruptcy was always a clear danger in the event of even limited shortfalls in demand for output. See World Bank, 'Report and Recommendations', op. cit., p. 20.

65. One analyst reported the symptoms of conspicuous consumption to include the building of ostentatious homes in Seoul and the use of Mercedes and Cadillacs. The behavior of this élite prompted a poet, Kim Chi-ha, to publish a poem criticizing the military generals, leading industrialists, prominent bureaucrats, members of the cabinet and representatives of the National Assembly, whom he referred to as the 'Five Thieves'. This poem prompted a government crackdown on the journal in which it was published and the prosecution of all who had been involved, including the editors and staff writers and the poet himself. The government prosecuted the group for its intention to stir up 'class struggle' which was in violation of the anti-communist law. The area of Seoul, where expensive houses of government officials and industrialists are centered is now known as 'Five Thieves Village'. See Joungwon A. Kim, op. cit., p. 279.

66. As the World Bank report described it:

> The allocational efficiency of the system of directed [government controlled] lending has been satisfactory in terms of the aggregate flow of funds towards the priority areas of heavy industry. It appears, however, that the magnitude of subsidies has resulted in excess demand for debt financing and may have led to some misallocation of funds within the *too broadly defined priority areas*. Favored access to

financing for ill-defined priority borrowers may have led to suboptimal investment decisions with respect to product choice, scale of production, and capital intensity. Excessive credit expansion has also fueled inflation and has encouraged a high level of indebtedness in the corporate sector. The high indebtedness of Korean industry appears to be worrisome owing particularly to the high proportion of short term debt resulting from easy access to cheap export credits. (World Bank, 'Report and Recommendations', op. cit., p. 26.)

See also Ho, 'South Korea and Taiwan', op. cit., p. 1182.

67. This sector is defined to include: (a) general plant equipment; (b) power plant equipment; (c) heavy electrical equipment; (d) marine diesel engines; (e) heavy industrial equipment and related components. In plant equipment, over $1 billion has been allocated since the mid-1970s. In heavy electrical equipment, another $200 million was invested, while the value of investment in marine diesel engines is estimated to have been over $200 million since the mid-1970s.

68. This tortuous effort to rationalize a duplicative industrial structure did result in the government's acceptance of one basic principle: that the Korean local market would be unable to support more than one plant equipment manufacturer. However, this cost the government more resources in subsidies in order to create a new firm. Similar measures have been attempted in other areas of this sector – illustratively, the government forced the merger of Sangyong and Korong into the Hyosung company, which was then designated sole producer of heavy electrical equipment.

69. See Henry Scott Stokes, 'South Korea under Chun: a New Sense of Vigor', *New York Times*, 4 March 1982, p. 6. In May 1982, disclosure of a multimillion-dollar loan swindle by a relative of Chun's wife put the credibility of Chun's régime in temporary jeopardy. According to recent analysis, however, Chun's prompt actions – which included changes in cabinet and government posts – helped to stymie opposition to Chun himself and to deflect public outcry. It is nevertheless true that rumors about continuing financial corruption in government are still a part of public perception in Korea. See Tracy Dahlby, 'Chun's Actions Ease Loan Scandal Impact', *Washington Post*, 23 May, 1982, p. 5.

70. Vreeland *et al.*, op. cit., pp. 179–80.

71. The officials who are commonly described as having influence in the Park régime include the deputy prime minister, who served also as the minister of the superagency, the Economic Planning Board (whose formal powers and political support were enormous), Prime Minister Kim Jong-pil; the ministers of defense, commerce and industry, and finance; and two or three economic secretaries on the staff of the Blue House. In addition, the heads of the KCIA, Ministry of Home Affairs (which controlled the national police), Capital Garrison Command and Army Security Command were considered significant in the power structure. See Olsen, op. cit., p. 72; Se-Jin Kim and Chi-won Kang, op. cit., pp. 153–8; and Vreeland *et al.*, op. cit., p. 180.

72. This is the view of an American diplomat who served for years in the US Embassy, Seoul.
73. There are many reasons why migration to, and high population density in Seoul may remain a fairly immutable problem. Seoul has been the capital of Korea for 600 years; in a highly centralized and culturally homogenous country, it has always been the case that social success and status can be attained only in the capital city. Added to this is the fact that Korea's economic growth has been closely managed by the government, especially with regard to the provision of special privileges to favored firms. This has made business location in Seoul a top priority for success. Seoul also has 60 per cent of all colleges and universities. It is considered impossible to really succeed professionally unless one is educated in a Seoul University. (From interviews in Seoul, March 1982.)

    The ratio of housing units to households has declined steadily from 82.7 per cent in 1960 to 74.4 in 1976. The low rate of investment in housing is directly attributed by the World Bank to 'deliberate government decision to emphasize investment in facilities for industrialization'. World Bank, 'Korea: Rapid Growth and Search for New Perspectives', unpublished, report no. 2477–KO (15 May 1979) p. ix.
74. Korea relies on imports for 65 per cent of its energy needs and 100 per cent of its requirements for petroleum. See World Bank, 'Korea: Current Developments and Policy Issues', report no. 30005-KO (20 May 1980) p. 23, and Samuel P. S. Ho, 'South Korea and Taiwan', op. cit., p. 1185. See also World Bank, 'Report and Recommendations', op. cit., p. 23.
75. Samuel P. S. Ho, 'South Korea and Taiwan', op. cit., p. 1183, and World Bank, 'Report and Recommendations', op. cit., pp. 1–2.

Chapter Five: Conclusion: a Broad View of Security

1. The nationalistic resonances of the concept of self-reliance were not lost on the Nixon administration. As we have seen in earlier chapters, the urge toward self-reliance, or autonomy, is native to the normal, healthy state. The manifestations of this urge were clear enough at the time President Nixon introduced his doctrine.
2. The National Taiwan Institute of Technology, which was established in 1974, provided advanced training for technical careers. It was projected to have enrolled over 2000 students by 1980. See Ralph N. Clough, *Island China* (Cambridge, Mass.: Harvard University Press, 1978) pp. 120–31. Moreover, there has been an effort to use the military in Taiwan for development since the 1960s. The burden of a 600,000-man military inevitably caused strains on the labor supply, especially in earlier years in the rural areas. Aside from the use of the military for construction and 'civic action' work in the rural sector, a technical vocational training program was added to the military training program for draftees in this period. See Hungdah Chiu, 'The Future of Political Stability in Taiwan', in US Congress, Senate, Committee on Foreign Relations, *Taiwan: One Year After United States–China Normalization* (Washington, DC: US

Government Printing Office, 1980) p. 37. As a measure of the bias toward the liberal arts among Taiwan's students, Chiu estimates that in 1978–9 about 40 per cent of the island's students were studying the humanities, fine arts, social sciences, and the law (ibid., p. 42).

3. Wanki Paik, 'Psychocultural Approach to the Study of Korean Bureaucracy', in Se-Jin Kim and Chi-won Kang, *Korea: a Nation in Transition* (Seoul: Research Center for Peace and Unification, 1971) p. 241.

4. Tong-Hui Lee, 'The Outlook of the Fifth Republic of Korea,' unpublished (Korean Military Academy, 28 July 1982) p. 2.

5. James C. Davies, 'The J-Curve of Rising and Declining Satisfactions as a Cause of Some Great Revolutions and a Contained Rebellion', in Hugh Davis Graham and Ted Robert Gurr, *The History of Violence in America* (New York: Bantam Books, 1969), cited in Chong-Sik Lee, 'South Korea 1979: Confrontation, Assassination and Transition', *Asian Survey*, xx, 1 (January 1980) 66. Labor discontent has been an important source of political instability in Korea, in spite of strict government prohibitions against independent labor activities. There are only government-sanctioned labor unions in Korea, which operate within tight parameters set by government decree. Unauthorized labor activity is generally dealt with very harshly (ibid., p. 76).

6. Gregory Henderson summed up this particular Korean political dynamic: 'South Korea seems likely to continue with a democratically tinged authoritarianism, publicly tolerated while successful, unstable when beleaguered.' See Henderson, *Korea: the Politics of the Vortex* (Cambridge, Mass.: Harvard University Press, 1968) p. 367.

7. Beginning in the mid-1970s, President Park invested considerable effort and personal prestige in developing advanced missiles, particularly missiles with greater range than those which the United States had permitted to be transferred. Park perceived military advantages to be derived from the possession of a missile capable of hitting Pyong-yang, a military objective that US defense advisors considered neither necessary nor desirable. (From interviews in Seoul, 1982.) Much attention was paid during this period to Park's concurrent efforts to develop nuclear capabilities, undoubtedly related in a significant way to the missile program. These are thought to have been terminated in 1975–6 as a result of US pressure. See, for example, Robert Gillette, 'US Squelched Apparent South Korea A-Bomb Drive', *Los Angeles Times*, 4 November 1978, p. 1; 'South Korea Suspension Last Week of Negotiations with France', *Nucleonics Week* (5 February 1976) pp. 9–10; Young-sun Ha, 'Nuclearization of Small States and World Order: the Case of Korea', *Asian Survey*, xi (November 1978) 1134–51. Although the nuclear development program was reported in the United States to have been under the aegis of a special *ad hoc* government committee that reported directly to the Blue House (the so-called Weapons Exploitation Committee), the effort to develop a missile delivery capability for nuclear as well as conventional warheads was apparently under the control of the Agency for Defense Development, the R&D arm of the Ministry of Defense. It is thought that pressure from President Park on its director to achieve results led to severe upheavals within ADD. In the end, ADD consistently exaggerated the progress it was making in order

to maintain its position with Park. Although estimates of the costs of this program are not available, they are thought to have been extremely high, causing severe constraints on other development goals. (From interviews in Seoul, 1982.)

8. Given the considerable ambiguity contained in the August 1982 US communiqué concerning continued US sales of armaments to Taiwan, it does appear that the US will continue to provide defensive equipment, in keeping with the statutes of the Taiwan Relations Act. In addition, Taiwanese engineers and scientists continue to be trained in the United States. While they must be engaged in ostensibly civilian pursuits in order to receive US approval, many of them are gaining experience in activities useful for military programs.

9. Some Koreans have argued that sales to Iraq are consistent with US foreign policy, given shared antipathy to Iran. Efforts by Iran to acquire armaments from Korea before the Gulf War were squelched – in part by US pressure – which led to the downgrading of Korean-Iranian relations. Koreans have pointed to the Iranian example as an instance of Korean restraint in arms sales and an indicator of cooperation with US policy, ignoring the fact that the US played a major part in forestalling the contracts. (From interviews in Seoul, 1982.)

10. A recent example of defense innovation in a country with developing defense industries, Israel, provides a good example of what the combination of incentives and technical capabilities can produce. Although publicly available evidence is still limited, there are indications that Israel's defense industry has achieved breakthroughs in anti-tank capabilities. The devastation of Syrian tanks by Israeli anti-armor weapons during the 1982 war in Lebanon was achieved on a scale unprecedented in prior conflicts, suggesting to some that Israel had even mastered a complex technology known as sense-and-destroy-armor still in the testing stage in the United States. See Jonathan C. Randal, 'Israel Said to Use Super Weapon in Destroying Syrian Tanks', *Washington Post*, 1 July 1982. It is more likely that the weapon in question was the relatively advanced cluster munition, Rockeye, whose capabilities previously may have been underestimated.

11. There are several underlying reasons why the army remained the most important service in Taiwan. Some stem from internal factors, some from deliberate US policy. Firstly, the army was the senior service in the 1950s, and the largest. As such, it not only enjoyed greater political influence but had a self-perpetuating quality. It was considered easier in the 1950s to retain the large number of senior army personnel than to find for them occupations in what was then still a very limited economic structure. Secondly, the commitment to the defense of the offshore islands of Quemoy and Matsu required far larger ground forces than would have been required for the defense of Taiwan (and Pescadores) alone. Thirdly, the United States emphasized extensive armaments for Taiwan's army, along the lines of 'limited offensive capabilities', as a contingency in the event that Taiwan might be called upon to assist the United States in a far larger war effort in East Asia. By contrast, the Taiwanese navy and air force were seen in 1950 simply as adjuncts to the US Seventh Fleet. See Clough, *Island China*, op. cit., pp. 105–7.

12. A good example of the ascendancy of pragmatic leaders was the appointment of Admiral Soong Chang-chin as Minister of Defense in November 1981.
13. For a fuller discussion of the structure of defense in South Korea, see Ralph Clough, *Deterrence and Defense in South Korea* (Washington, DC: Brookings Institution, 1978).

Bibliography

Abramowitz, Morton, 'Moving the Glacier: the Two Koreas and the Great Powers', *Adelphi Paper*, no. 80 (London: International Institute for Strategic Studies, Spring 1977).

Agapos, A. M., *Government – Industry and Defense: Economics and Administration* (University of Alabama Press, 1975).

Albrecht, U., Ernst, D., Lock, P., and Wulf, H., 'Militarization, Arms Transfer and Arms Production in Peripheral Countries', *Journal of Peace Research*, XII (1975) 195–212.

Alexander, Arthur, 'Economic Motivations in International Arms Production and Transfers: an Integrated Framework to Analyze Policy Alternatives' (Harvard University, 16 May 1980).

Almond, Gabriel A., and Verba, Sidney, *The Civic Culture* (Princeton University Press, 1963).

Amden, Alice, 'Taiwan's Economic History', *Modern China* (July 1979) 341–80.

American Institute in Taiwan, 'Foreign Economic Trends and their Implications for the United States', Draft (December 1981).

American Institute in Taiwan, cable on Taiwanese New Ten Year Development Plan (Taipei, December 1981).

Anderson, Jack, 'A Bizarre Interpretation of Taiwan Issue', *Washington Post*, 26 January 1982, p.B15.

'As American Reps See It: ROC Stressing Hi-technology', *Free China Review* (March 1982) 36.

Ashley, Fred, US Department of State, Washington DC, interview, April 1982.

Asia Report, II, 12 (December 1981) 2–3.

Balaschak, Mark, Ruina, Jack, Steinberg, Gerald, and Yaron, Anselm, 'Assessing the Comparability of Dual-use Technologies for Ballistic Missile Development', report no. ACOWC113 prepared for the US Arms Control and Disarmament Agency (June 1981).

Baldwin, Frank (ed.), *Without Parallel: the American–Korean Relationship since 1945* (New York: Random House, 1974).

Ball, Nicole, *The Military in the Development Process* (Claremont, Ca.: Regina Books, 1981).

Ball, Nicole, 'The Military in Politics: Who Benefits and How', *World Development*, IX, 6 (1981) 569–83.

Ball, Nicole, 'Defense and Development: a Critique of the Benoit Study' unpublished (Swedish Institute of International Affairs, May 1982).

Barclay, George W., *Colonial Development and Population in Taiwan* (Princeton University Press, 1954).

Barnds, William J., *The Two Koreas in East Asian Affairs* (New York University Press, 1976).

Barnett, A. Doak, *Communist China and Asia* (Washington, DC: Brookings Institution, 1952).

Barnett, A. Doak, *China and the Major Powers in East Asia* (Washington, DC: Brookings Institution, 1977).

Barnett, A. Doak, *US Arms Sales: the China–Taiwan Tangle* (Washington, DC: Brookings Institution, 1982).

Barton, John H., and Imai, Ryukichi (eds), *Arms Control II* (Cambridge, Mass.: Oelgeschlager, Gunn and Hain, 1981).

Benjamin, Milton R., 'South Korea Doubles Atomic Power Plans', *Washington Post*, 13 November 1978, p.1.

Benoit, Emile, 'Growth Effects of Defense in Developing Economies', *International Development Review*, I (1972).

Benoit, Emile, with Millikan, Max E., and Hagen, Everett E., *Effect of Defense on Developing Economies*, 2 vols, Arms Control and Disarmament Agency report no. E–136 (Cambridge, Mass.: Center for International Studies, Massachusetts Institute of Technology, June 1971).

Berney, Karen, 'Dual-use Technology Sales', *China Business Review* (July–August 1980) 23–6.

Bertrand, Harold E., *The Defense Industrial Base*, report no. SD–321, prepared for US Department of Defense (Washington, DC: Logistics Management Institute, 1977).

Blechman, Barry M. and Berman, Robert P. (eds), *Guide to Far Eastern Navies* (Annapolis: Naval Institute Press, 1978).

Bloch, Peter C., 'Is Democratic Government a Luxury for Developing Countries?', paper presented to the International Security Studies Conference, Fletcher School of Law and Diplomacy, Boston, Mass. (24–6 April 1978).

Browning, E. S., 'East Asian Economies', *Foreign Affairs*, LX, 1 (Autumn 1981) 123–47.

Brzoska, Michael, and Wulf, Herbert, 'Rejoinder to Benoit's "Growth and Defense in Developing Countries" – Misleading Results and Questionable Methods', paper from the study group on Armaments and Underdevelopment, University of Hamburg (1979).

Buchanan, Norman, and Ellis, Howard S., *Approaches to Economic Development* (New York: Twentieth Century Fund, 1955).

Bureau of National Affairs, *Federal Contracts Report: Foreign Military Sales: a Review and Some Thoughts* (Washington, DC: Bureau of National Affairs, January 1981).

Canby, Steven L., and Luttwak, Edward N., 'The Control of Arms Transfers and Perceived Security Needs', report prepared for the US Arms Control and Disarmament Agency, ACDA contract no. AC9WC112 (Washington, DC: C & L Associates, June 1979).

Chang, King-Yuh, 'Partnership in Transition: a Review of Recent Taipei–Washington Relations', *Asian Survey*, XXI, 6 (June 1981) 603–21.

Chang, King-Yuh (ed.), *Western Pacific Security in a Changing Context: Problems and Prospects* (Taipei: Freedom Council, 1982).

Chapman, William, 'Brown in Seoul to Reassert Commitment', *Washington Post*, 7 November 1978, p.14.

Chapman, William, 'Seoul to Get More US Arms', *Washington Post*, 19 October 1979, p.1.

Chapman, William, 'Korean Talks Deadlock Amid Angry Exchanges on Alleged Provocations', *Washington Post*, 6 April 1980, p.21.

Chapman, William, 'Ex-Im Bank to Continue Seoul's Loans Despite Military Rule', *Washington Post*, 6 June 1980, p.16.

Chapman, William, 'Taiwan's Leaders Expect Reagan to Upgrade Ties', *Washington Post*, 5 July 1980, p.1.

'China Says It's Willing to Negotiate on Issue of US Arms Sales to Taiwan', *Baltimore Sun*, 1 February 1982, p.1.

Chiu Hungdah (ed.), *China and the Question of Taiwan: Documents and Analysis* (New York: Praeger, 1973).

Chiu, Hungdah (ed.), *China and the Taiwan Issue* (New York: Praeger, 1979).

Chiu, Hungdah, 'Military Preparedness and Security Needs: Perceptions from the Republic of China on Taiwan', *Asian Survey*, XXI, 6 (June 1981).

Chodak, Szymon, *Societal Development* (New York: Oxford University Press, 1973).

Choi, Hochin, 'The Process of Industrial Modernization in Korea: the Latter Part of the Chosun Dynasty through the 1960s', *Journal of Social Sciences and Humanities*, XXVI (June 1967) 1–33.

Cline, Ray S., 'Peking Will Press Reagan for Changes', *New York Times*, 6 January 1981, p.22.

Clough, Ralph N., *East Aisa and US Security* (Washington, DC: Brookings Institution, 1975).

Clough, Ralph N., *Deterrence and Defense in Korea* (Washington, DC: Brookings Institution, 1976).

Clough, Ralph N., *Island China* (Cambridge, Mass.: Harvard University Press, 1978).

Coffey, Raymond, 'Soviets Use US Technology to Build Arms, Official Says', *Chicago Tribune*, 20 November 1979, p.3.

Cole, D. C., and Lyman, P. N., *Korean Development – the Interplay of Politics and Economics* (Cambridge, Mass.: Harvard University Press, 1971).

Copley, Gregory, 'Third World Arms Production', *Defense and Foreign Affairs Digest* (September 1978) 26–41.

Curasi, Richard H., 'Neither Puppet nor Pariah Be: the Development of the South Korean Defense Industry as an Indicator of the Quest for Independence, Self-reliance, and Sovereignty', unpublished paper prepared for the Naval Post-graduate School, Monterey, California (September 1979).

Dahlby, Tracy, 'Ancient Enmities Cast Shadow on South Korea's Ties with Japan', *Washington Post*, 29 January 1982, p.23.

Dahlby, Tracy, 'Chun's Actions Ease Loan Scandal Impact', *Washington Post*, 23 May 1982, p.5.

Dahlby, Tracy, 'North, South Shadowbox Across Korea's Iron Triangle', *Washington Post*, 4 June 1982, p.21.

Defense and Foreign Affairs Handbook, 1981 (New York: Franklin Watts, 1981).

Defense Market Survey, Inc., *DMS Market Intelligence Report for China (Taiwan) and South Korea* (Greenwich, Conn.: Defense Market Survey, Inc., 1981).

'Defense Production in the Third World', *Ground Defense, Armies and Weapons*, VII, 58 (November 1979).

Deshingkar, G. D., 'The Arms Race: a Perspective on Asia', *Alternatives*, V, 2 (August 1979) 253–73.

Domes, Jurgen, 'Political Differentiation in Taiwan: Group Formation within the Ruling Party and the Opposition Circles 1979–1980, *Asian Survey*, XXI, 10 (October 1981) 1011–28.

Downen, Robert L., *Of Grave Concern: US–Taiwan Relations on the Threshold of the 1980s* (Washington, DC: Center for Strategic and International Studies, Significant Issues Series, 1981).

Dudzinsky, S. J., and Digby, James, 'Qualitative Constraints on Conventional Armaments', report prepared for the US Arms Control and Disarmament Agency, contract no. R–1957–ACDA (Santa Monica, Calif.: Rand Corporation, July 1976).

'Economic Development = National Security', *Far Eastern Economic Review*, 18 May 1979, pp.52–4.

Edwards, John, 'The Sino-American Embrace', *Far Eastern Economic Review*, 1 August 1980, pp.32–4.

Eiland, Lt. Col. Michael D., 'Effects of Military Expenditures on ASEAN Economies', State Department memorandum (2 December 1977).

Evans, Rowland, and Novak, Robert, 'Taiwan Turnabout', *Washington Post*, 18 January 1982, p.21.

'Export Controls: Dual-use Focus', *China Business Review* (January–February 1980) 6–7.

'F–5 Design at Twenty-five', *Air Force Magazine* (April 1982) 42.

Faini, Ricardo, Arnez, Patricia, and Taylor, Lance, 'Defense Spending, Economic Structure and Growth: Evidence Among Countries and Over Time', unpublished (Cambridge, Mass.: Massachusetts Institute of Technology, October 1980).

Fairbank, John King, *The United States and China* (Cambridge, Mass.: Harvard University Press, 3rd edn, 1972).

'Fear Led Seoul to Seek A-Weapons, Probers Say', *Baltimore Sun*, 2 November 1975, p.5.

Fei, John C. H., *et al.*, *Growth with Equity: The Taiwan Case*, World Bank Report 0-19-52-0116-7 (1976).

Fink, Donald E., 'Nationalists Update Fighter Force', *Aviation Week and Space Technology*, 29 May 1978.

Fink, Donald E., 'Aerospace on Taiwan-2 Center Designs Two Aircraft', *Aviation Week and Space Technology*, 5 June 1978.

Fitzgerald, Stephen, *China and the Overseas Chinese* (London: Cambridge University Press, 1972).

Flanders, June M. 'Prebisch on Protectionism: an Evaluation', *Economic Journal*, LXXIV, 294 (June 1964) 305–26.

Foster, James L., 'New Conventional Weapons Technologies: Implications for the Third World', paper prepared for the Conference on the Implications of the Military Build-up in Non-industrial States, Fletcher School of Law and Diplomacy, 6–8 May 1976.

Gail, Bridget, 'The "Fine Old Game of Killing", Comparing US and Soviet Arms Sales', *Armed Forces Journal International* (September 1978) 16–24.

Gail, Bridget, 'The Fine Old Game of Killing, Part Two', *Armed Forces Journal International* (November 1978) 37–9.

Galenson, Walter (ed.), *Economic Growth and Structural Change in Taiwan* (Ithaca & London: Cornell University Press, 1979).

Gansler, Jacques, *The Defense Industry* (Cambridge, Mass. & London: MIT Press, 1980).

Gelb, Leslie, 'US, Peking and Taipei Make Game of Arms Sales', *New York Times*, 16 October 1982, p.1

General Research Corporation, *The Impact on the Rationalization of European Defense Industry of Alternative US Approaches to Transatlantic Defense Cooperation* (McLean, Va.: GRC, April 1979).

George, Alexander L., 'Case Studies and Theory Development: the Method of Structured, Focused Comparison', in *Diplomatic History: New Approaches*, ed. Paul Gordon Lauren (New York: Free Press, 1979) pp.43–69.

Gervasi, Tom, *Arsenal of Democracy* (New York: Grove Press, 1977).

Gessert, Robert A., 'The Dependence of European Defense Industry on Arms Exports as a Problem of International Cooperation', summary of research proposal presented by the General Research Corporation to the US Arms Control and Disarmament Agency (January 1979).

Getler, Michael, 'New Military Strength in Asia', *Washington Post*, 5 April 1982, p.20.

Getler, Michael, 'US Moves to Sell Aircraft Parts to Taiwan', *Washington Post*, 14 April 1982, p.1.

Gibert, Stephen P., *Northeast Asia in US Foreign Policy*, Washington Papers vol. VII (Beverly Hills & London: Sage Publications, Georgetown Center for Strategic and International Studies, 1979).

Gibney, Frank, 'The Ripple Effect in Korea', *Foreign Affairs* (October 1977) 160–74.

Gillette, Robert, 'US Squelched Apparent South Korea A-Bomb Drive', *Los Angeles Times*, 4 November 1978, p.1.

Goldstein, Dr Donald J., 'Third World Arms Industies: their Own Slings and Swords', unpublished paper.

'Government to Spur Cleanup to Achieve a Just Society', *Korea News Review*, 12 December 1981, pp.5–6.

Greene, Fred, *US Policy and the Security of Asia* (New York: McGraw-Hill, 1968).

Griffith, David R., 'Sales Data Cleared for China', *Aviation Week and Space Technology*, 9 June 1980, pp.16–17.

Griffith, William E., *Peking, Moscow and Beyond* (Beverly Hills: Sage Publications, 1973) pp.1–5.

Groth, Carl, 'The Economics of Weapons Coproduction', paper presented to the International Studies Association Conference, Los Angeles, 20 March 1980.

Ha, Young-sun, 'Nuclearization of Small States and World Order: the Case of Korea', *Asian Survey*, XI (November 1978) 1134–51.

Halloran, Richard, 'New Jets for Taiwan: an Issue Surrounded by Nettles', *New York Times*, 27 January 1981, p.2.

Halloran, Richard, 'Reagan and Peking Aide Skirt Arms Issues in Talks', *New York Times*, 30 October 1981, p.1.

Halloran, Richard, 'Taiwan Jet Sale Near Approval, Officials Report', *New York Times*, 10 November 1981, p.1.

Halloran, Richard, 'Weinberger to Press Koreans' Rights and Support for GIs in 3-Day Visit', *New York Times*, 29 March 1982, p.4.

Halloran, Richard, 'Weinberger Says US Will Maintain Curbs on Seoul's Sale of Arms', *New York Times*, 1 April 1982, p. 5.

Halperin, M. H., *The 1958 Taiwan Straits Crisis: a Documented History*, RM 2900 ISA (Santa Monica, Calif.: Rand Corporation, December 1966).

Hammond, Paul Y., Louscher, David J., and Salomon, Michael D., 'Controlling US Arms Transfers: the Emerging System', *Orbis* (Summer 1979) 317–35.

Harding, Harry Jr, *China and the US: Normalization and Beyond* (New York: China Council, 1979).

Henderson, Gregory, *Korea: the Politics of the Vortex* (Cambridge, Mass.: Harvard University Press, 1968).

Henderson , Gregory, Lebow, Richard Ned, and Stoessinger, John G., *Divided Nations in a Divided World* (New York: David McKay, 1975).

Higgins, Benjamin, *Economic Development: Principles, Problems and Policies* (New York: W. W. Norton, 1959).

Ho, Samuel P. S., 'Economic Development of Taiwan 1860–1970', manuscript (New Haven 1977).

Ho, Samuel P. S., 'South Korea and Taiwan: Development Prospects and Problems in the 1980s', *Asian Survey*, XXI, 12 (December 1981) 1175–97.

Holland, Max, 'The Myth of Arms Restraint', *International Policy Report* (Washington, DC: Center for International Policy, May 1979).

Hollie, Pamela G., 'Northrop's Strategy: Simplify', *New York Times*, 4 February 1980, p.1.

Hsiao, Frank S. T., and Sullivan, Lawrence R., 'The Politics of Reunification: Beijing's Initiative on Taiwan', *Asian Survey*, XX, 8 (August 1980) 769–802.

Hsiao, Frank S. T., Sullivan, Lawrence R., and Horowitz, Irving L., 'Military Origins of the Cuban Revolution', *Armed Forces and Society*, 4 (1975) 41–8.

Hu, Lt. Gen. Yu-Tung, Ministry of National Defense, Taipei, Taiwan, interview, March 1982.

Huntington, Samuel P., *The Common Defense* (New York: Columbia University Press, 1961).

Huntington, Samuel P., *Political Order in Changing Societies* (New Haven: Yale University Press, 1964).

Huntington, Samuel P., and Moore, Clement H., *Authoritarian Politics in Modern Society* (New York: Basic Books, 1970).

Hwang, Dong Joon, Korea Institute for Defense Analysis, Seoul, Korea, interview, March 1982.

International Institute for Strategic Studies, *The Military Balance 1981–1982* (London 1981).

Jacobs, J. Bruce, 'Taiwan 1978: Economic Successes, International Uncertainties', *Asian Survey*, XIX, 1 (January 1979) 21–9.

Jacobs, J. Bruce, 'Taiwan 1979: "Normalcy" After "Normalization" ', *Asian Survey*, XX, 1 (January 1980) 84–93.

Jacobs, J. Bruce, 'Political Opposition and Taiwan's Political Future', *Australian Journal of Chinese Affairs*, 6 (1981) 21–44.

Jacoby, Neil H., *US Aid to Taiwan* (New York: Praeger, 1966).

Janowitz, Morris, *The Military in the Political Development of New Nations* (University of Chicago Press, 1964).

Jo, Yung-Hwan (ed.), *Taiwan's Future?* (Tempe, Ariz.: Center for Asian Studies, Arizona State University, 1974).

Johnson, John J. (ed.), *The Role of the Military in Underdeveloped Countries* (Princeton University Press, 1962).

Johnson, Stuart E., and Yager, Joseph A., *The Military Equation in Northeast Asia* (Washington, DC: Brookings Institution, 1979).

Jordan, Amos, and Taylor, William J., *American National Security* (Baltimore & London: Johns Hopkins University Press, 1981).

Jorgenson, Dale W., 'The Development of a Dual-economy', *Economic Journal*, LXXI (June 1961) 309–34.

Kamm, Henry, 'In Taipei, No One Wants Peking Ties', *New York Times*, 16 October 1980, p.9.

Kearns, Kevin, US Embassy, Seoul, Korea, interview, March 1982.

Keatley, Robert, 'China Makes Overture to Taiwan', *Wall Street Journal*, 1 October 1981, p.6.

Kennedy, Gavin, *The Military in the Third World* (London: Gerald Duckworth, 1974).

Keon, Michael, *Korean Phoenix: a Nation from the Ashes* (Englewood Cliffs, N. J.: Prentice-Hall, 1977).

Khil, Young Whan, and Bark, Dong Suh, 'Food Policies in a Rapidly Developing Country: the Case of South Korea, 1960–1978', *Journal of Developing Areas*, 16 (October 1981) 47–70.

Kilpatrick, James J., 'Taiwan: Things Have Changed', *Washington Post*, 2 September 1982, p.24.

Kim, Chang K., Patton and Morgan Corporation, New York, interview, February 1982.

Kim, Chongwhi, National Defense College, Seoul, Korea, interview, March 1982.

Kim, Joungwon A., *Divided Korea: the Politics of Development 1945–1972* (Cambridge, Mass.: Harvard University Press, 1975).

Kim, Oh-Sung, Poong-San Metal Corporation, Seoul, Korea, interview, March 1982.

Kim, Se-Jin, 'South Korea's Involvement in Vietnam and its Political and Economic Impact', *Asian Survey*, X, 6 (June 1970).

Kim, Se-Jin, *The Politics of Military Revolution in Korea* (Chapel Hill: University of North Carolina Press, 1971).

Kim, Se-Jin, and Kang, Chi-Won, *Korea: a Nation in Transition* (Seoul: Research Center for Peace and Unification, 1971).

Kim, Youn-soo, 'The ROK, the DPRK and Yugoslavia: 1950–1978', *Korea and World Affairs*, II, 2 (Summer 1978) 218–46.

Kindermann, Gootfried-Karl, 'Washington between Beijing and Taipei: the Restructured Triangle 1978–1980', *Asian Survey*, XX, 5 (May 1980) 457–77.

Kindleberger, Charles P., *Economic Development* (New York: McGraw–Hill, 1958).

King, John Jerry (ed.), *International Political Effects of the Spread of Nuclear Weapons* (Washington, DC: US Government Printing Office, April 1979).

Klare, Michael T., and Holland, Max, 'Conventional Arms Restraint: an Unfulfilled Promise', special report of the Coalition for a New Foreign and Military Policy, the Institute for Policy Studies and the Center for International Policy, Washington, DC (1979).

Klass, Phillip J., 'E2C Radar to Provide New Flexibility', *Aviation Week and Space Technology*, 12 July 1976, pp.51–3.

Knowles, Wallace, US Department of Defense, interview, April 1982.

Korea Development Institute, *Long-term Prospects for Economic and Social Development, 1977–1991* (Seoul: Korea Development Institute, 1978).

'Korea, US No Longer See Eye to Eye', *Baltimore Sun*, 2 July 1982, p.2.

'Korean Defense Production', *Ground Defense/Armies and Weapons* (November 1979) 5–6.

Krueger, Anne O., *The Development Role of the Foreign Sector and Aid* (Cambridge, Mass.: Harvard University Press, 1979).

Kurata, Phil, 'Violence in the Name of Reason', *Far Eastern Economic Review*, 28 December 1979, p.27.

Kuznets, Paul W., *Economic Growth and Structure in the Republic of Korea* (New Haven & London: Yale University Press, 1977).

Kwon, Soon Tae, Ministry of Foreign Affairs, Seoul, Korea, interview, March 1982.

Kyle, Joe, American Institute in Taiwan, Washington, DC, interview, April 1982.

Lachica, Eduardo, 'Proposed Arms Sales to China, Taiwan Promise Political, Not Military, Dispute', *Wall Street Journal*, 5 October 1981, p.32.

LaPalombra, J. (ed.), *Bureaucracy and Political Development* (Princeton University Press, 1963).

Lasserre, Philippe, and Boisot, Max, 'Strategies and Practices of Transfer of Technology from European to ASEAN Enterprises', unpublished report prepared for the European Economic Community by Institut Européen d'Administration des Affaires, Euro-Asia Center, Fontainebleau, France (April 1980).

Lee, Chong-Sik, 'South Korea 1979: Confrontation, Assassination and Transition', *Asian Survey*, XX, 1 (January 1980) 63–76.

Lee, Chong-Sik, 'South Korea in 1980: the Emergence of a New Authoritarian Order', *Asian Survey*, XXI, 1 (January 1981) 125–43.

Lee, George, American Institute in Taiwan, Taipei, Taiwan, interview, March 1982.

Lee, Tong-Hui, 'The Outlook of the Fifth Republic of Korea', unpublished (Korean Military Acadamy, 28 July 1982).

Leitenberg, Milton, 'The Military Implications of Arms Sales to the Third

World', paper presented to the Conference on International Arms Trade, Institute for Policy Studies, Washington, DC (22–3 April 1978).

Lenorowitz, Jeffrey M., 'Taiwan Technology Outlook Bright', *Aviation Week and Space Technology*, 11 June 1979, pp. 145–6.

Leontieff, W., and Hoffenberg, M., 'The Economic Impact of Disarmament', *Scientific American* (April 1961).

Lewis, John Wilson, 'China's Military Doctrine and Force Posture', in *China's Quest for Independence: Policy Evolution in the 1980s*, ed. Thomas F. Fingar and the Stanford Journal of International Studies (Boulder: Westview Special Studies, 1979).

Li, Victor H. (ed.), *The Future of Taiwan: a Difference of Opinion* (White Plains, N. Y.: M. E. Sharpe, 1980).

Lilly, James, American Institute in Taiwan, Taipei, Taiwan, interview, March 1982.

Lin, Ching-yuan, *Industrialization in Taiwan, 1946–1972* (New York: Praeger, 1973) chs 5–7.

Liu, Melinda, 'Taiwan Develops Nuclear Industry', *Washington Post*, 27 February 1977, p.21.

Liu, Melinda, 'Accounting for the N-Factor', *Far Eastern Economic Review*, 17 December 1978, pp. 22–4.

Lock, Peter, and Wulf, Herbert, *Register of Arms Production in Developing Countries* (Hamburg: Study Group on Armaments and Underdevelopment, 1977).

Lohr, Steve, 'South Korea Betting on Ships', *New York Times*, 5 June 1982, p.29.

Louscher, David J., 'The Rise of Military Sales as a US Foreign Assistance Instrument', *Orbis* (Winter 1977) 933–64.

McGowan, Pat and Kegley, Charles (eds.) *Threats, Weapons and Foreign Policy* (Beverly Hills, Calif.: Sage Publications, 1980) p.186.

Mann, Paul, 'Military Growth Spurs Economic Debate', *Aviation Week and Space Technology*, 30 November 1981, pp.117–24.

Mann, Paul, 'China Export Policy Takes Final Form', *Aviation Week and Space Technology*, 25 January 1982, pp.57–8.

Mason, Edward S., Kim, Mahn Je, Perkins, Dwight H., Kim, Kwang Suk, and Cole, David C., *The Economic and Social Modernization of the Republic of Korea* (Cambridge, Mass. & London: Harvard University Press, 1980).

Meade, Grant E., *American Military Government in Korea* (New York: King's Crown Press, Columbia University, 1951).

Meier, Gerald M., *The International Economics of Development* (New York: Harper & Row, 1968).

Melman, Seymour, *The Defense Economy: Conversion of Industries and Occupations to Civilian Needs* (New York: Praeger, 1972).

Meyers, Ramon H., 'The Economic Development of Taiwan' in *China and the Question of Taiwan: Documents and Analysis* (New York: Praeger, 1973).

Miller, Stephen E., 'Arms and the Third World: the Indigenous Weapons Production Phenomenon', unpublished paper prepared for the Programme for Strategic and International Affairs, Graduate Institute of International Studies, University of Geneva (June 1980).

Mills, Major Charles J., Joint US Military Assistance Group-Korea, interview, March 1982.

Misch, Franz, US Embassy, Seoul, Korea, interview, March 1982.

Moodie, Michael, *Sovereignty, Security and Arms*, vol. 7, no. 67 of *The Washington Papers* (Beverly Hills & London: Sage Publications, 1979).

Moodie, Michael, 'Vulcan's New Forge: Defense Production in Less Industrialized Countries', *Arms Control Today*, x, 3 (March 1980).

Moran, Theodore H., *Multinational Corporations and the Politics of Dependence: Copper in Chile* (Princeton University Press, 1974).

Mossberg, Walter S., 'Pentagon Resists Japan Computer Sale to China', *Wall Street Journal*, 22 September 1981, p.1.

Mossberg, Walter S., 'US to Push Less Capable F–X Jet Fighters for Sale to 11 Major Third World Allies', *Wall Street Journal*, 10 August 1982, p.6.

Myrdal, Gunnar, *Rich Lands and Poor* (New York: Harper & Row, 1957).

Neuman, Stephanie G., 'Security, Military Expenditures and Socioeconomic Development: Reflections on Iran', *Orbis* (Fall 1978) 588–92.

Neuman, Stephanie G., and Harkavy, Robert E., *Arms Transfers in the Modern World* (New York: Praeger, 1979).

Niksch, Larry A., 'US Troop Withdrawal from South Korea: Past Shortcomings and Future Prospects', *Asian Survey*, xxi, 3 (March 1981) 326–41.

Northrop Corporation, 'Fighter Coproduction and Security Assistance', presentation document on the F–5G program, NB81–95 (May 1981).

Novak, Jeremiah, 'China, Taiwan's Suitor', *New York Times*, 7 October 1981.

'Nunn Favors Talks on Direct US Arms Sales to China', *Aerospace Daily*, 8 January 1980, p.2.

Oberdorfer, Don, 'Carter Rejects Plan on Early F–16 Sale to Korea', *Washington Post*, 25 January 1978, p.17.

Oberdorfer Don, 'North Korea's Army Now Ranked Fifth Largest in World by US', *Washington Post*, 14 January 1979, p.9.

Oberdorfer, Don, 'US Troop Pullout in Korea Dropped', *Washington Post*, 21 July 1979, p.3.

Oberdorfer, Don, 'Absence of Peking Arms Buyers May Be Hint to US on Taiwan', *Washington Post*, 18 September 1981, p.1.

Oberdorfer, Don, 'China, US Fail to Get Agreement on Taiwan Arms', *Washington Post*, 31 October 1981, p.13.

Oberdorfer, Don, 'US Reported to Reject Jet Sale to Taiwan', *Washington Post*, 11 January 1982, p.1.

Oberdorfer, Don, 'US Discloses Barring Sale of Warplanes to Taiwan', *Washington Post*, 12 January 1982, p.5.

Oberdorfer, Don, 'Arms Sales to Taiwan Will Be Sent to Hill', *Washington Post*, 28 March 1982, p.19.

Oksenberg, Michael, 'China Policy for the 1980s', *Foreign Affairs* (Winter 1980/1981) 302–22.

Oliver, Richard, 'Employment Effects of Reduced Defense Spending', *Monthly Labor Review* (June 1971).

Olsen, Edward A., '"Korea, Inc.": The Political Impact of Park Chung Hee's Economic Miracle', *Orbis* (Spring 1980) 69–84.

Parisi, Anthony J., 'Seoul May Purchase Reactors in Europe', *Washington Post*, 2 March 1978, p.1.

Park, Chung Hee, *Our Nation's Path: Ideology of Social Reconstruction* (Seoul, Korea: Hollym Corporation, 1970).

Park, Eul Y., Korea Development Institute, Seoul, Korea, interview, March 1982.

Park, Tong Whan, 'The Korean Arms Race: Implications in the International Policies of Northeast Asia', *Asian Survey*, XX, 6 (June 1980).

Pauker, Guy J., Canby, Steven, Johnson, A. Ross, and Quandt, William B., *In Search of Self-reliance: US Security Assistance to the Third World under the Nixon Doctrine*, report prepared for the Advanced Research Projects Agency, grant no. R–1092–ARPA (Santa Monica, Calif.: Rand Corporation, June 1973).

'Peking Presses Detente with the Taiwan Regime', *New York Times*, 4 October 1981, p.6.

Peleg, Ilan, 'Military Production in Third World Countries: a Political Study', in *Threats, Weapons and Foreign Policy*, ed. Pat McGowan and Charles W. Kegley, Jr (Beverly Hills & London: Sage Publications, 1979).

Perkins, Capt. Malcolm H., Joint US Military Advisory Group-Korea, Seoul, Korea interview, March 1982.

Perkins, Malcolm H., and Bolles, Wilhelm, 'Security Assistance to South Korea: Assessment of Political, Economic and Military Issues from 1975 to 1979', Masters thesis, Air Force Institute of Technology, Air University, Wright-Patterson Air Force Base, Dayton, Ohio (1979).

Pfaltzgraff, Robert L., Jr, 'China, Soviet Strategy and American Policy', *International Security*, V, 2 (Fall 1980).

Pierre, Andrew, *The Global Politics of Arms Sales* (Princeton University Press, 1982).

Pollack, Johnathan D., 'China's Potential as a World Power', *Internationl Journal*, XXXV, 3 (Summer 1980) 580–95.

Poong-San Metal Corporation, 'Defense Industry in Korea', internal memorandum (March 1982).

Pound, Edward T., 'Curbs on Technology Exports Hurt by Gaps in Enforcement', *New York Times*, 14 October 1981, p.1.

'Pratt & Whitney, Korea discuss F100 Coproduction', *Aerospace Daily*, 4 June 1979, p.165.

Prebisch, Raul, 'Commercial Policy in the Underdeveloped Countries', *American Economic Review, Papers and Proceedings*, XLIV (May 1959) 251–73.

Prebisch, Raul, *Towards a New Trade Policy for Development*, United Nations Report on the Conference on Trade and Development (New York, 1964).

Pye, Lucian W., 'Dilemmas for America in China's Modernization', *International Security*, IV, 1 (Summer 1979) 3–20.

Ra'anan, Uri, Pfaltzgraff, Robert L., and Kemp, Geoffrey, *Arms Transfers to the Third World: the Military Buildup in Less Industrial Countries* (Boulder: Westview Press, 1976).

Randal, Jonathan C., 'Israelis Said to Use Super Weapon in Destroying Syrian Tanks', *Washington Post*, 1 July 1982, p.22.

Ravenal, Earl, 'Consequences of the End Game in Vietnam', *Foreign Affairs*, 53 (July 1975) 114–43.

188 Bibliography

Reeve, W. D., *The Republic of Korea: a Political and Economic Study* (London: Oxford University Press, 1963).

Republic of China (Taiwan), Council for Economic Planning and Development, *Ten-year Economic Development Plan for Taiwan, Republic of China 1980-1989* (1980).

Republic of China (Taiwan), Council for Economic Planning and Development, *Taiwan Statistical Yearbook* (1981).

Republic of China (Taiwan), Economic Advisory Office, *Economic Development in the Republic of China* (Taipei, 1979).

Republic of China (Taiwan), Ministry of Economic Affairs, *Economic Indicators: Taiwan, Republic of China* (Taipei, 1978).

Republic of China (Taiwan), Ministry of Education, *Educational Statistics of the Republic of China* (Taipei, 1981).

Republic of China, Ministry of Education, *Education in the Republic of China* (Taipei, 1981)

Republic of Korea, Korean Overseas Information Sevice, Ministry of Culture and Information, *A Handbook of Korea* (Seoul, Korea: Samhwa Printing Company, 1979).

Republic of Korea, Economic Planning Board, *A Summary Draft of the Fifth Five-year Economic and Social Development Plan 1982–1986* (September 1981).

Robinson, Clarence A., Jr, 'Benha Pushes Defense, Civil Technology', *Aviation Week and Space Technology*, 25 January 1982, pp. 59–61.

Ropelewski, Robert, 'F-16/79 Offers Export Market New Capability', *Aviation Week and Space Technology*, 23 November 1981, pp. 38–47.

Ropka, Lawrence, Jr, American Institute in Taiwan, Taipei, Taiwan, interview, March 1982.

Ross, A. L., 'Conventional Arms Production in Developing Countries', paper presented to the International Studies Association Conference, Los Angeles, California (March 1980).

Rostow, W. W., 'The Take-off into Self-sustained Growth', *Economic Journal* (March 1956) 422–63.

Rupp, Rainer W., 'China's Strategic Aims and Problems of Military Modernization', *NATO Review* (February 1981) 14–20.

Ryu, Chung, Poong-San Metal Corporation, Seoul, Korea, interview, March 1982.

' "Saemaul" Worthy Cause', *Korea News Review*, 20 March 1982.

Scalapino, Robert A., 'Asia at the End of the 1970s', *Foreign Affairs, America and the World, 1979*, 58, 3 (1980) pp. 693–738.

Scalapino, Robert A., *The United States and Korea: Looking Ahead* (Beverly Hills & London: Sage Publications, 1979).

Scalapino, Robert A., 'Korean Dynamics', *Problems of Communism*, xxx (November-December 1981) 16–32.

Schumacher, E. F., *Small is Beautiful: Economics as if People Mattered* (New York: Harper & Row, 1973).

Science Applications, Inc., *Technology List for Observing Possible Indigenous Development/Production of a Surface-to-Surface Missile System by a Less Developed Country (LDC)*, report prepared for the US Arms Control and Disarmament Agency, no. AC8WC122 (Arlington, Va., April 1979).

Senia, Al, 'Northrop Takes a $300 Million Gamble', *Washington Post*, 8 August 1982, p.1.

'Shipping in Asia', *Far Eastern Economic Review*, 28 February 1975, p.9.

Singer, W. H., 'The Distribution of Gains Between Investing and Borrowing Countries', *American Economic Review, Papers and Proceedings* (May 1950).

Smith, Bruce A., 'Koreans Seek New Military Air Capacity', *Aviation Week and Space Technology*, 22 October 1979, pp.62–3.

Smith, M. T., 'US Foreign Military Sales: its Legal Requirements, Procedures and Problems', paper presented to the Conference on Implications of the Military Build-up in Non-industrial States, Fletcher School of Law and Diplomacy, Boston, Mass. (6–8 May 1976).

Snyder, Edwin K., Gregor, A. James, and Chang, Maria Hsia, *The Taiwan Relations Act and the Defense of the Republic of China* (Berkeley: Institute of International Studies, University of California, 1980).

Social Action Coordinating Committee, 'Brief on the Inertial Technology Training Program at MIT', unpublished report on training of Taiwanese students at MIT (1976).

Solomon, Richard H., 'Thinking Through the China Problem', *Foreign Affairs* (January 1978) 324–56.

Solomon, Richard H. (ed.), *Asian Security in the 1980s: Problems and Policies for a Time of Transition*, report no. R–249215A prepared for the Office of the Assistant Secretary of Defense, International Security Affairs (Santa Monica, Calif.: Rand Corporation, November 1979).

Solomon, Richard H. (ed.), *The China Factor* (Englewood Cliffs: Prentice-Hall, 1981).

'South Korea, Pakistan Sign for F–16s', *Aerospace Daily*, 23 December 1981, p.278.

'South Korea Suspension Last Week of Negotiations with France', *Nucleonics Week*, 5 February 1976, pp.9–10.

'South Korea's Arms Industry: Boom-Boom', *Economist*, 2 December 1978, p.36.

'Spies in the Classroom', *Newsweek*, 17 May 1982, p.73.

Steinberg, Gerald M., 'The Cost of Independence: the Economic Impacts of the Israeli Defense Industry', in *The Role of Defense Industries in the Industrial Structures of Modern Nations*, ed. Milton Leitenberg and Nicole Ball (forthcoming).

Sterba, James P., 'Taiwan Denounces Overtures by Peking for Talks' *New York Times*, 16 September 1981, p.14.

Sterba, James P., 'Taiwan Rejects as Propaganda Peking's Offer of Talks', *New York Times*, 10 October 1981.

Sternheimer, Stephen, 'The Impact of Differentiated Technology Transfer from the West to the People's Republic of China: Political, Social, Economic, and Military Implications and US Policy Options', report prepared for the US Department of Defense, US DR&E contract no. MDA 903–80–C–0275 (1 September 1980).

Stockholm International Peace Research Institute, *Nuclear Energy and Nuclear Weapon Proliferation* (London: Taylor & Francis, 1979).

Stockholm International Peace Research Institute, *World Armaments and Disarmament: SIPRI Yearbook* (London: Taylor & Francis, annual).

Stokes, Henry Scott, 'Cracking Down in Korea', *New York Times Magazine*, 19 October 1980, p. 110.

Stokes, Henry Scott, 'Taiwan's Premier Hopes Reagan Sends New Arms', *New York Times*, 25 January 1981, p.11.

Stokes, Henry Scott, 'South Korea under Chun: a New Sense of Vigor', *New York Times*, 4 March 1982, p. 6.

Stokes, Henry Scott, 'Anti-US Sentiment is Seen in Korea', *New York Times*, 28 March 1982, p.2.

Stokes, Henry Scott, 'Negotiations on Japanese Aid to Seoul Seem Near Impasse', *New York Times*, 19 April 1982, p.6.

Sunoo, Harold Hakwon, *America's Dilemma in Asia: the Case of South Korea* (Chicago: Nelson-Hall. 1979).

Sutter, Robert G., and Mills, William deB., 'Fighter Aircraft Sales in Taiwan: US Policy', Congressional Research Service, issue brief no. IB81157, 28 October 1981 (updated 20 January 1982).

Taylor, John H., 'The F–5E Tiger II: a "Sports" Model with Punch', *Air Force Magazine* (January 1975) 55–57.

'Text of Statement by Peking', *New York Times*, 1 October 1981, p.12.

The Security of Northeast Asia in the 1980s: National Perspectives, working papers of the Northeast Asia–United States Forum's Strategic Issues Study Group, Stanford, Calif. (November 1980).

Tien, Hung–mao, 'Taiwan in Transition: Prospects for Socio–Political Change', *China Quarterly*, 64 (December 1975) 626.

Tugwell, Franklin, *The Politics of Oil in Venezuela* (Stanford University Press, 1975).

'Two Women, Man Sought in Pusan Arson Case', *Korea Times*, 20 March 1980, p.8.

Ulsamer, Edgar, 'Military Electronics', *Air Force Magazine*, 58 (July 1975) 42–51.

United Nations, Economic and Social Council, Committee for Development Planning, *Disarmament and Development: an Analytical Survey and Pointers for Action*, paper prepared for the Committee for Development Planning by Barry M. Blechman and Edward R. Fried (26 January 1977).

United Nations Association of the United States of America, *Beyond Normalization: Report of the UNA–USA National Policy Panel to Study US–China Relations* (New York, 1979).

US Arms Control and Disarmament Agency, *The International Transfer of Conventional Arms*, report to Congress pursuant to Section 302 of the Foreign Relations Authorization Act of 1972 (Washington, DC: US Government Printing Office, 1974).

US Arms Control and Disarmament Agency, *World Military Expenditures and Arms Transfers, 1969–1979*, ACDA Publication 112 (March 1982).

US Bureau of Labor Statistics, *Projections of a Post-Vietnam Economy, 1975* (Washington, DC: US Government Printing Office, 1972).

US Central Intelligence Agency, *Korea: the Economic Race Between the North and the South*, National Foreign Assessment Center Research Paper no. ER 78–10008 (January 1978).

US Central Intelligence Agency, *Arms Flows to LDCs: US–Soviet Comparisons*, National Foreign Assessment Center no. ER 78–10494U (November 1978).

US Congress, House, Committee on Foreign Affairs, *The International Transfer of Conventional Arms*, report from the US Arms Control and Disarmament Agency (Washington, DC: US Government Printing Office, 1974).

US Congress, House, Committee on Foreign Affairs, *Asia in a New Era: Implications for Future US Policy* (Washington, DC: US Government Printing Office, 1975).

US Congress, House, Committee on Foreign Affairs, *United States–Soviet Union–China: the Great Power Triangle*, 95th Congress, 1st session (1977)

US Congress, House, Committee on Foreign Affairs, *Department on Export Control of United States Technology*, 95th Congress, 1st session (1977).

US Congress, House, Committee on Foreign Affairs, *Review of the President's Arms Transfer Policy*, hearings before the Subcommittee on International Security and Scientific Affairs, 95th Congress, 2nd session (1978).

US Congress, House, Committee on Foreign Affairs, *International Transfer of Technology: an Agenda of National Security Issues* (Washington, DC: US Government Printing Office, 1978).

US Congress, House, Committee on Foreign Affairs, *Hearing Before the Subcommitee on International Operations on H.R. 11548*, 95th Congress, 2nd session (1978).

US Congress, House, Committee on Foreign Affairs, *Security Issues: Korea and Thailand*, 96th Congress, 1st session (1979).

US Congress, House, Subcommittee on International Security and Scientific Affairs, *Conventional Arms Transfer Policy* (Washington, DC: US Government Printing Office, 1978).

US Congress, House, Subcommittee on Asian and Pacific Affairs, *China and Asia – An Analysis of China's Recent Policy Toward Neighboring States* (Washington, DC: US Government Printing Office, 1979).

US Congress, House, Subcommittee on Asian and Pacific Affairs, *Playing the China Card: Implications for United States–Soviet–Chinese Relations*, report prepared by the Congressional Research Service (Washington, DC: US Government Printing Office, 1979).

US Congress, Senate, Committee on Appropriations, *Foreign Assistance and Related Program Appropriations, Fiscal Year 1975*, 93rd Congress, 2nd session (1974).

US Congress, Senate, Committee on Foreign Relations, *US Arms Sales Policy*, 94th Congress, 2nd session (1977).

US Congress, Senate, Committee on Foreign Relations, *US Troop Withdrawal from the Republic of Korea*, Senators Hubert H. Humphrey and John Glenn (Washington, DC: US Government Printing Office, 1978).

US Congress, Senate, Committee on Foreign Relations, *Prospects for Multilateral Arms Export Restraint* (Washington, DC: US Government Printing Office, 1979).

US Congress, Senate, Committee on Foreign Relations, *The United States, China and Japan* (Washington, DC: US Government Printing Office, 1979).

US Congress, Senate, Committee on Foreign Relations, *Sino–American*

Relations: a New Turn (Washington, DC: US Government Printing Office, 1979).

US Congress, Senate, Committee on Foreign Relations, *Taiwan: Hearings on US Policy in East Asia*, 96th Congress, 2nd session (1979).

US Congress, Senate, Committee on Foreign Relations, *Implementation of the Taiwan Relations Act: the First Year* (Washington, DC: US Government Printing Office, 1980).

US Congress, Senate, Committee on Foreign Relations, *Taiwan: One Year After United States–China Normalization* (Washington, DC: US Government Printing Office, 1980).

US Congress, Senate, Committee on Foreign Relations, *The Implications of US–China Military Cooperation* (Washington, DC: US Government Printing Office, 1981).

US Congress, Joint Economic Committee, *Economic Performance and the Military Burden in the Soviet Union* (Washington, DC: US Government Printing Office, 1970).

US Congress, Office of Technology Assessment, *Technology and East–West Trade* (Washington, DC: US Government Printing Office, 1979).

US Department of the Army, 'Processing of Export Licenses', unpublished memorandum.

US Department of Defense, Office of the Director of Defense Research and Engineering, *An Analysis of Export Control of US Technology – A DOD Perspective*, report of the Defense Science Board Task Force on Export of US Technology (4 February 1976).

US Department of Defense, US Design, Research and Engineering, *Recommendations: Multinational Codevelopment/Coproduction Workshop*, workshop held at Fort Belvoir, Defense Management College (21–4 October 1980).

US Department of Defense, *Charter for Review of DOD Policy Toward Coproduction/Industrial Participation Agreements*, memorandum (20 July 1981).

US Department of State, Munitions Control Newsletters (1976–82).

US Department of State, *Arms Transfer Policy*, 95th Congress, 1st session (1977).

US Department of State, *Report to Congress on Arms Transfer Policy*, 95th Congress, 2nd session (1977).

US Department of State, 'Standard Operating Procedure for Interdepartmental Clearance of Arms Transfers', internal memorandum from the Under Secretary of State for Security Assistance (24 January 1978).

US Department of State, *Transcript of Daily News Briefing* (2 February 1978).

US Department of State, Letter to Congressman Anthony C. Heilenson from Assistant Secretary for Congressional Relations, J. Brian Atwood, concerning the transfer of missile technology to South Korea (13 November 1979).

US Department of State, *Review of Arms Transfer Policy*, Current Policy Series, no. 145 (1980).

US Department of State, *US Position in the Pacific in 1980*, Current Policy Series, no. 154 (1980).

US Department of State, *China and the US: into the 1980s*, Current Policy Series, no. 187 (1980).

US Department of State, *Review of Relations with Taiwan*, Current Policy Series, no. 190 (1980).

US Department of State, *Taiwan*, Background Notes Series, no. 7791 (1980).

US Department of State, *FY1982 Proposals for Security Assistance*, Current Policy Series, no. 266 (1981).

US Department of State, 'Major Defense Equipment', *Munitions Control Newsletter*, 78 (January 1980).

US Department of State, Letter from Under Secretary Warren Christopher to Senator Frank Church, Chairman, Senate Foreign Relations Committee, concerning sale of advanced aircraft to Taiwan (11 July 1980).

US Department of State, 'Conventional Arms Transfer Policy', *Munitions Control Newsletter*, 90 (1981).

US Department of State, 'Industrial Outlook Report: Electronic Products', Airgram no. A–97 from US Embassy, Seoul (28 December 1981).

US Embassy, Seoul, 'Korean Government Support for Export Industries', Airgram no. A–81 (13 November 1981).

US Embassy, Seoul, *Economic Trends Report* (24 November 1981).

US Embassy, Seoul, 'Annual Labor Report for Korea', Airgram no. A–84 (November 1981).

US Embassy, Seoul, 'ROK'S Approved 1982 Central Government Budget and 1981 Supplemental Budget', Airgram no. A–93 (16 December 1981).

US Embassy, Seoul, 'Industrial Outlook Report: Electronic Products', Airgram no. A–97 (28 December 1981).

US Embassy, Seoul, 'Korean Mechandise Trade in 1981', Cable no. 01153 (3 January 1982).

US Embassy, Seoul, 'The 1982 Korean Economic Management Plan', Cable no. 15248 (3 January 1982).

US Embassy, Seoul, *Korea's Real GNP Increased by 7.1 per cent in 1981* (21 January 1982).

US Embassy, Seoul, 'Foreign Equity Investment in Korea, 1981', Airgram no. A–19 (5 March 1982).

US Embassy, Seoul, 'Korea's Balance of Payment: Results of 1981 and Plan for 1982', Airgram no. A–22 (16 March 1982).

US Embassy, Seoul, 'Revision of ROK's Major Tax Laws', Airgram no. A–13 (2 May 1982).

US Embassy, Taipei, 'Taiwan's Defense Industries', Cable no. 07721 (10 November 1978).

US General Accounting Office, *Coproduction Programs and Licensing Arrangements in Foreign Countries* (Washingon, DC: US Government Printing Office, 1975).

US General Accounting Office, *US Munitions Export Controls Need Improvement* (Washington, DC: US Government Printing Office, 1979).

US Library of Congress, Congressional Research Service, *Implications of President Carter's Conventional Arms Transfer Policy*, CRS no. 77–223F (Washington, DC, 22 September 1977).

US Library of Congress, Congressional Research Service, *China–US Relations*, Issue Brief no. IB76053 (1981).

US Library of Congress, Congressional Research Service, *Fighter Aircraft*

Sales to Taiwan: US Policy, Congressional Research Service Major Issues, Series no. IB81157, Robert G. Sutter and William deB. Mills (1981).

US President, Public Announcement of the Conventional Arms Transfer Policy, 19 May 1977.

US President, Statement, 'Arms Transfer Ceiling', *Federal Register* (2 February 1978).

'US Rice Dealers' Brawl', *Korea News Review*, 20 March 1982.

Vaitsos, Constantine V., *Intercountry Income Distribution and Transnational Enterprises* (Oxford: Clarendon Press, 1974).

Vargo, John J., Jr, Joint US Military Advisory Group-Korea, Seoul, interview, March 1982.

Vendevanter, Gen. E., 'Coordinated Weapons Production in NATO: a Study of Alliance Processes', contract no. RN–5282–PR (Santa Monica: Rand Corporation, 1967).

Verba, Sydney, 'Some Dilemmas in Comparative Research', *World Politics*, 20 (October 1976) 111–18.

Vreeland, Nena *et al.*, *Area Handbook for South Korea* (Washington, DC: US Government Printing Office, 1975).

Wachter, Robert, Northrop Corporation, Arlington, Virginia, interview, April 1982.

Walker, Ambassador Richard, US Embassy, Seoul, interview, March 1982

Wang, Chi-wu, 'Military Preparedness and Security Needs: Perceptions from the Republic of China on Taiwan', *Asian Survey*, XXI, 6 (June 1981) 650–68.

Wang, Chi-wu, 'Taiwan's Defense Policy in the Context of her Economic Development', unpublished paper prepared for the Conference on Security and Development in the Indo-Pacific, Fletcher School of Law and Dipolmacy, Boston, Mass. (24–26 April 1976).

Wang, Chi-wu, National Science Council, Taipei, interview, March 1982.

Wang, Sang-Eun, National Assembly, Seoul, interview, March 1982.

We, Yuan-li, *US Policy and Strategic Interests in the Western Pacific* (New York: Crane, Russak, 1975).

Weaver, J. L., 'Assessing the Impact of Military Rule: Alternative Approaches', in *Military Rule in Latin America*, ed. Phillipe Schmitter (London: Sage Publications, 1973).

Wei, Yung, 'Modernization Process in Taiwan: an Allocative Analysis', *Asian Survey*, XVI, 3 (March 1976), 24–69.

Wei, Yung, 'The Changing Security Environment in Asian Nations: Challenges and Opportunities', *Asian Pespective*, 2 (Autumn 1978) 134–49.

Weinraub, Bernard, 'Opposition Growing on Korean Pullout', *Washington Post*, 20 January 1979, p.12.

Weinstein, Franklin B., and Kamiya, Fuji, *The Security of Korea: US and Japanese Perspectives on the 1980s* (Boulder, Col.: Westview Press, 1980).

Weisskopf, Michael, 'China Delays Start of Talks on Arms Sales', *Washington Post*, 12 August 1981, p.3.

Weisskopf, Michael, 'Chinese Take a Great Leap Upward in Plane Resembling a Boeing', *Washington Post*, 11 December 1981, p.37.

Weisskopf, Michael, 'Taiwan Finds it Takes Two to Tango', *Washington Post*, 10 January 1982, p.2.

Weisskopf, Michael, 'China Hits US on Taipei Arms Sales', *Washington Post*, 12 January 1982, p.1.

Wells, Louis T. (ed.), *The Product Life Cycle and International Trade* (Boston: Division of Research, Harvard Graduate School of Business Administration 1972).

West, Robert L., 'Impact of International Economic Instability upon Security and Development', paper prepared for the Conference on Security and Development in the Indo-Pacific, Fletcher School of Law and Diplomacy, Boston, Mass. (24–6 April 1978).

West, Robert L., 'Economic Performance of Indo-Pacific Developing Countries 1967–1975', report prepared for the International Security Studies Conference, Fletcher School of Law and Diplomacy, Boston, Mass. (24–6 April 1978).

Wilson, George C., 'Services Told to Push New Warplanes', *Washington Post*, 7 August 1982, p.1.

Wolpin, Miles D., 'Socialism and Civilian Supremacy vs. Militarism in the Third World: a Comparison of Development Costs and Benefits', paper prepared for the Annual Meeting of the Canadian Science Association, University of New Brunswick, Fredericton, N.B. (9–11 June 1977).

World Bank, *'Korea: Rapid Growth and Search for New Perspectives'*, unpublished, report no. 2477–KO, East Asia and Pacific Regional Office (15 May 1979).

World Bank, *Staff Appraisal Report: Korea Sector Program on Higher Technical Education*, East Asia and Pacific Regional Office (11 January 1980).

World Bank, 'Korea: Current Developments and Policy Issues', report no. 30005–KO, East Asia and Pacific Regional Office (20 May 1980).

World Bank, *Korea: Staff Appraisal Report: the Korea Long-term Credit Bank, the Korea Development Finance Corporation VIII*, East Asia and Pacific Regional Office (20 November 1980).

World Bank, *World Development Report 1981* (Washington, DC August 1981).

World Bank, 'Report and Recommendations of the President of the International Bank for Reconstruction and Development to the Executive Directors on a Structural Adjustment Loan to the Republic of Korea', unpublished, report no. P–3156–KO (17 November 1981).

World Bank, *Korea Adjusting to a New World Environment*, East Asia and Pacific Regional Office (1 June 1982).

Wren, Christopher S., 'Taiwan Arms Issue Damaging US–Chinese Ties', *New York Times*, 28 February 1982, p.18.

Wright, Edward Reynolds (ed.), *Korean Politics in Transition* (Seattle & London: University of Washington Press, 1975).

Ying, Diane, 'Two Leading US Aircraft Builders Gunning for Lucrative Taiwan Sale', *Asian Wall Street Journal*, 21 July 1980, p.8.

Zagoria, Donald S., 'Why We Can't Leave Korea', *New York Times Magazine*, 2 October 1977.

Index